MOLECULAR
AND
ATOMIC SPECTROSCOPY

MOLECULAR
AND
ATOMIC SPECTROSCOPY

R. Wilfred Sugumar

Head
Department of Chemistry
Madras Christian College Chennai

MJP
PUBLISHERS

Chennai   Trichy   Tirunelveli   New Delhi

ISBN 978-81-8094-036-1   **MJP PUBLISHERS**

All Rights Reserved   No. 44 Nallathambi Street,
Printed And Bound In India   Triplicane, Chennai 600 005
MJP 224   © Publishers, 2019

Publisher : C. Janarthanan

This book has been published in good faith that the work of the author is original. All efforts have been taken to make the material error-free. However, the author and publisher disclaim responsibility for any inadvertent errors.

To

My mother

PREFACE

This Book is a humble attempt to introduce the rudiments of various spectroscopic techniques, which are widely used in modern analytical procedures. Recent developments in analytical chemistry have resulted in virtual explosion of techniques with respect to the types of samples that can be analysed and the sensitivity of analyte detection. Sophisticated studies of air and water quality, contamination of the food chain and effects of trace elements on the properties of solid-state devices require analytical methods that have been developed only recently. The present scenario in the analytical field demands the understanding of both the principles and the basic instrumentation involved in each spectroscopic technique. This text provides a readable and yet sound introduction to contemporary analytical methods with a focus on spectroscopic methods and it covers most of the common spectroscopic methods used in analytical chemistry.

This book has eleven chapters and all of them are written taking into account this view. In every chapter apart from explaining the fundamentals of conventional techniques, care is taken to include materials on currents trends. This includes techniques like FT-IR, 2D NMR, COSY,GC-MS, MALDI. A separate chapter on the chemical analysis of surfaces is included and this is a very emerging field particularly in the area of nanotechnology. This chapter describes the application of various methods such as surface spectroscopy, secondary ion mass spectrometry and Auger emission spectroscopy. Each chapter has been made as self-explanatory as possible and therefore it is not necessary to read a major part of the book in order to understand any particular chapter.

Enough examples have been included in the text to make the reading more interesting. References at the end of the book will lead the students to more detailed and advanced discussions of the methods selected.

I should record here my sincere thanks for the support I derived from my family members. My wife Ponni and my daughter Anita checked the text and urged me to be simple in my writing and they contributed greatly to the readability of the book.

I would like to acknowledge the helpful comments and suggestions of many students and colleagues. I also take this opportunity to acknowledge my thanks and appreciation to MJP Publishers, Chennai. They tolerated the delay on my part and brought this book with great dedication.

I truly appreciate comments and suggestions from students and teachers, which would help us in improving the book in its future editions.

R. WILFRED SUGUMAR

CONTENTS

Chapter 1

INTRODUCTION TO SPECTROSCOPY

Spectroscopy, Spectrometry and Spectrum

Measuring the absorption and emission of electromagnetic radiation is a key function in analytical chemistry. We use light in quantitative experiments to determine chemical concentrations, and we use light in a qualitative way to detect compounds. **Spectroscopy** is a science that deals with transitions in a chemical species induced by its interaction with the photons of electromagnetic radiation. Spectroscopy involves the investigation of interaction of electromagnetic radiation with matter as well as the explanation of spectral patterns. The plot of absorption or transmittance of electromagnetic radiation as a function of wavelength (or wave number) is called **spectrum**. The measurement of absorption and emission of electromagnetic radiation by materials is called **spectrophotometry** or spectrometry, in short. Spectrophotometry uses electromagnetic radiation to measure chemical concentrations.

The Electromagnetic Spectrum—Spectrometry From Radio Frequency To Gamma Rays

When we say *light*, we usually mean visible light. However, the visible spectrum constitutes a very small part of the total radiation spectrum. Most of the radiation cannot be seen, but can be detected by instruments. This **electromagnetic spectrum** ranges from very short wavelengths (e.g. gamma and X-rays) to very long wavelengths (e.g. microwaves and broadcast radio waves). Figure 1.1 shows many of the important regions in electromagnetic spectrum and demonstrates the inverse relationship between wavelength and frequency.

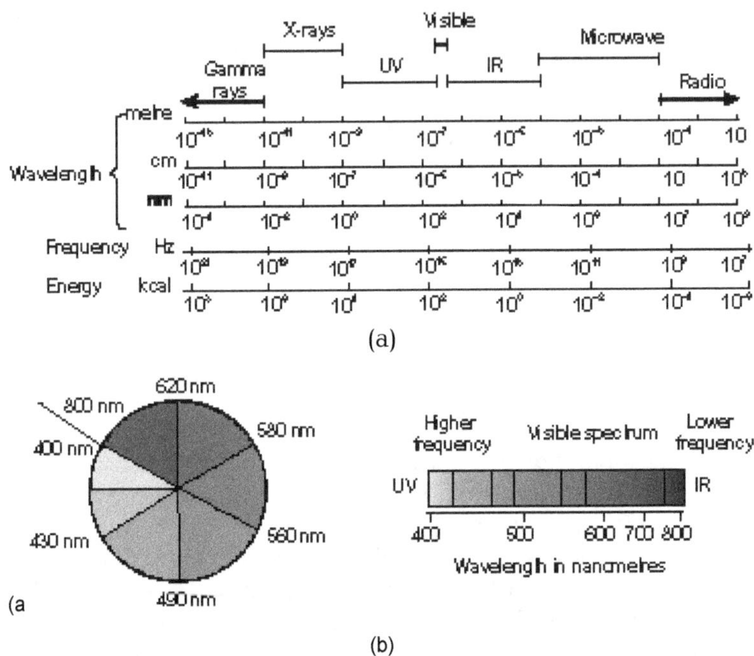

Figure 1.1 (a) Electromagnetic spectrum (b) Visible spectrum

Review of Energy, Wavelength, Frequency and Temperature

Light can be described both as waves and as particles. The light waves consist of perpendicularly oscillating electric and magnetic fields. The **wavelength**, λ, is the crest-to-crest distance between waves (Figure 1.2). The **frequency**, ν, is the number of complete oscillations that the waves make each second. The unit of frequency is reciprocal seconds, i.e., s^{-1}. One oscillation per second may also be called one **hertz** (Hz). A frequency of 10^6 s^{-1} is, therefore, said to be 10^6 Hz, or one **megahertz** (MHz). The product of frequency and wavelength is c, the speed of light (2.998×10^8 m/s in vacuum.

$$\nu\lambda = c \quad (11)$$

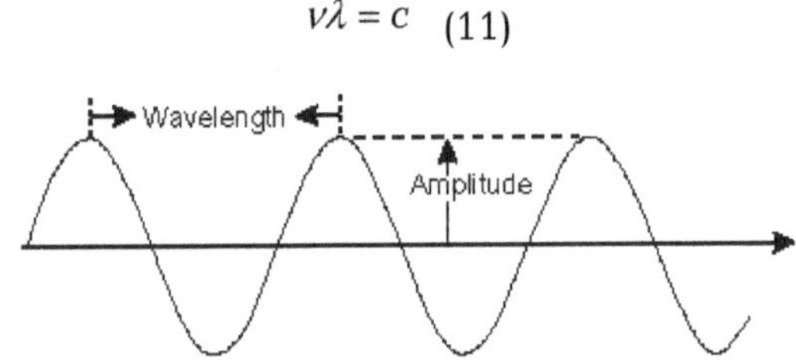

Figure 1.2 Wavelength and amplitude of light

Light can also exist as particles called **photons.** The energy, E, of a photon is proportional to its frequency.

$$E = h\nu \qquad (1.2)$$

where h is the Planck's constant (6.626×10^{-34} J s^{-1}).

Combining equations 1.1 and 1.2

$$E = h\frac{c}{\lambda} = hc\frac{1}{\lambda} = hc\bar{\nu} \qquad (1.3)$$

where $\bar{\nu} = \dfrac{1}{\lambda}$ is the **wavenumber**. Energy is inversely proportional to wavelength and directly proportional to wavenumber. Energy increases when frequency increases, wavelength decreases and wavenumber increases.

Temperature is a measure of average kinetic energy of an object. Since energy is closely related to wavelength and frequency, they both must be functions of temperature as well. Energy is directly proportional to temperature and can be expressed either as the energy per atom

$$E = K_B T$$

where K_B is Boltzmann's constant (1.308×10^{-16} erg K^{-1} atom^{-1} or 1.380×10^{-23} JK^{-1} atom^{-1}), or as the energy per mole of material, i.e.,

$$E = RT$$

where R is gas constant (8.3145×10^{7} erg K^{-1} mol^{-1} or 8.3145 J K^{-1} mol^{-1}).

Example 1

What is the wavelength of radiation of a microwave oven whose frequency is 2.45 GHz?

Solution

2.45 GHz means 2.45×10^{9} Hz = 2.45×10^{9} S^{-1}

$$\lambda = \frac{c}{\nu} = \frac{2.998 \times 10^{8}\,\text{m/s}}{2.45 \times 10^{9}\,s^{-1}} = 0.122\ \text{m}$$

Example 2

A molecule absorbs visible light with a wavelength of 500 nm. By how many joules is the energy of a molecule increased?

Solution

$$E = h\nu = hc/\lambda$$
$$= (6.626 \times 10^{-34}\,\text{Js}^{-1}) \times (2.998 \times 10^{8}\,\text{m/s}) / 500 \times 10^{-9}\,\text{m})$$
$$= 3.97 \times 10^{-19}\,\text{J}$$

Transformation of Light Energy

As per the law of conservation of energy, the sum of all forms of energy entering a material must equal the sum of all forms of energy leaving plus the energy remaining in the material. The energy is carried into and out of an atom as light, as heat or as kinetic energy of particles such as electrons. Light can be transformed into heat energy when the wavelength that excites the atom is less than the wavelength of light that is emitted. The spectrometry of the forms of energy that impinge on an atom and the forms that leave the atom are listed in Table 1.1.

Table 1.1 Transformation of light energy interacting with atoms

Energy in	Energy out	Spectrometry
Heat	Light	Emission
Light	Heat	Absorption
Light	Light	Luminescence
Light	Moving electrons	Photoelectron
Moving electrons	Moving electrons	Auger
Bonding energy	Light	Chemiluminescence

Measurement of Spectra

In spectrometry, the measurement of spectra depends both on the spectral properties of the sample and the instrument used for measurement. The following three fundamental components of a spectrometric instrument are required to carry out measurements in any region of the electromagnetic spectrum.

1. A source of electromagnetic radiation must be present.

2. The effects from a single wavelength of electromagnetic radiation must be obtained. This single wavelength may arise either from the source or from the emission of analyte itself.

3. A transducer must be present to measure changes in the quantity of electromagnetic radiation falling on it.

As much light as possible from a stable source is directed into a scanning monochromator. As light of continuously varying wavelengths is passed through the sample and through the exit slit on to a transducer, a graph of power versus time (calibrated as wavelength) results in spectrum.

In some cases, it is useful to consider that the molecule has the ability to absorb light even when the absorption is not being measured. The graph of incipient absorbance versus wavelength is called the absorption envelope.

Emission and Absorption Spectrometry

In emission spectrometry, the emission of the sample itself provides the light, and the intensity of light emitted is a function of the analyte concentration. A monochromator is placed between the emitting sample and the transducer. Emission instruments are used in atomic spectrometry as well as in a few assays involving chemiluminescence.

Absorption spectrometry involves measuring the fraction of light of a given wavelength that passes through a sample. The sample does not emit light by itself, and so a separate radiant source must be included. Most sources produce light of undesired wavelengths in addition to the ones desired. Passing light through either a monochromator or a filter selects the wavelength for the assay. Measurements are made when the analyte or calibration samples are in place.

The terms used in absorption spectrometry are listed in Table 1.2.

Table 1.2   Spectrophotometry nomenclature

Name	Symbol	Definition	Name not recommended
Absorbance	A	$-\log T$	Optical density, extinction absorbency
Absorptivity	A	$-A/bc$	Absorbency index, extinction coefficient
Path length	B	Internal cell or sample length in cm	l or d
Molar absorptivity	ε	$-A/bc$	Molar extinction coefficient, molar absorption coefficient
Transmittance	T	I/I_o	Transmittancy or transmission
Wavelength unit	nm	10^{-9} m	mμ (millimicron)
	μm	10^{-6} m	μ
Absorption maximum	λ_{max}	Wavelength at which maximum absorption occurs	

Absorptivity, A, is in grams per litre; molar absorptivity, ε, is in moles per litre

Fluorescence and Phosphorescence Spectrometry

In luminescence spectra, a light source for the appropriate energy region is required, and the incident light is monochromatic. The incident light arrives at the sample after passing through the first monochromator although monochromatic light sources such as laser can also be used. Luminescence that comes from the sample is measured at an angle that is not in line with the axis connecting the source and the sample. This angle is often, but not necessarily, 90°. If the sample luminescence is proportional to the analyte content, it can be used to quantify the analyte. If a time-dependent spectrometric method is used, the light due to phosphorescence and that due to fluorescence may be separated.

Diffuse Reflectance Spectrometry

Reflected light can be classified into two kinds. The first is the light that returns from a smooth surface. This is called **specular reflection**, which is usually seen when light reflects from a mirror or a flat interface where a change of index or refraction occurs. The second kind of reflected light is spread out in all directions either by rough interfaces between a solid and a gas or within a solid at the interfaces of crystallites. This scattering is called **diffuse reflection**. The amount of reflected light is quantitated as reflectance, which is the ratio of total reflected flux of light to the total incident flux. Quantitative measurements of diffuse reflectance are most precisely made by using an integrating sphere. It collects all the light that returns upward from a surface. Of the light collected, a constant fraction of that radiation reaches the detector. Experimentally, the reflectance ratio is found by comparing the spectrum of light reflected from samples with the spectrum reflected from standards that have high reflectivities over a wide range of wavelengths.

REVIEW QUESTIONS

1. Define

    i.   Spectroscopy

    ii. Spectrum

    iii. Spectrophotometry

2. What is a Hertz?

3. Distinguish between absorption and emission spectrometry.

4. Calculate the energy per photon (in ergs) of radiation with wavelength 380 nm.
    (Ans: 5.23×10^{-12} erg/photon)

5. Calculate the wavelength (in μm) of radiation with frequency 8.58×10^{13} sec^{-1}.
    (Ans: 3.50 μm)

6. In which region of electromagnetic spectrum do the following frequencies exist?

    i. 1000 cm^{-1} ii. 60000 cm^{-1} iii. 700 MHz   iv. 5 cm^{-1}

    (Ans: i. Infrared ii. Ultraviolet iii. Radio frequency iv. Microwave)

7. Discuss the important characteristics of electromagnetic radiation.

8. Calculate the wavenumber of a radiation of wavelength 4 μm.
    (Ans: 2500 cm^{-1})

9. The frequency of strong yellow line in the spectrum of sodium is 5.09×10^4 s^{-1}. Calculate the wavelength in nanometres.
    (Ans: 589.4 nm)

10. Convert 6000 A$^\circ$ into frequency units Hz.
    (Ans: 5×10^{14} Hz)

Chapter 2

ELECTRONIC SPECTROMETRY

Ultraviolet and Visible Spectroscopy

The difference in energy between molecular bonding, non-bonding and anti-bonding orbitals ranges from 125 to 650 kJ per mole. This energy corresponds to electromagnetic radiation in the ultraviolet region and visible region of the spectrum. Thus, UV–visible spectroscopy deals with the measurement of energy absorbed when electrons are promoted to higher energy levels. The UV–visible region of electromagnetic radiation is depicted in Figure 2.1.

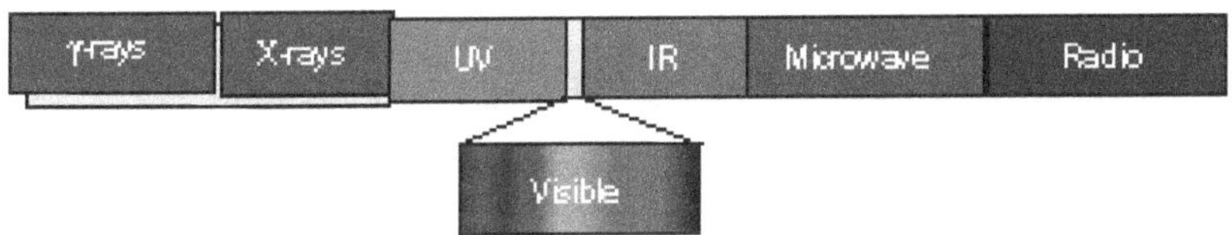

Figure 2.1 UV–visible region in electromagnetic spectrum

The UV–visi ble spectrum can be divided into three regions.

1. Far or vacuum ultraviolet region (10–200 nm)

2. Near or quartz ultraviolet region (200–380 nm)

3. Visible region (380–780 nm)

The absorption of visible light makes things colourful. For example, a blue dye used in a pair of jeans appears blue because the light at the red end of the spectrum is absorbed. This leaves the blue light to be reflected to the observer's eye (Figure 2.2).

UV–visible spectroscopy is useful as an analytical technique for two reasons. First, it can be used to identify some functional groups in molecules; secondly, it can be used for assaying. This second role—determining the content and strength of a substance—is extremely useful. UV–visible spectroscopy is used extensively in chemical and biochemical laboratories for a variety of tasks. For example, it can be used to determine even the trace metal content in an alloy or the amount of a certain drug reaching various parts of the body. The most ubiquitous use of UV is as a detection device for HPLC.

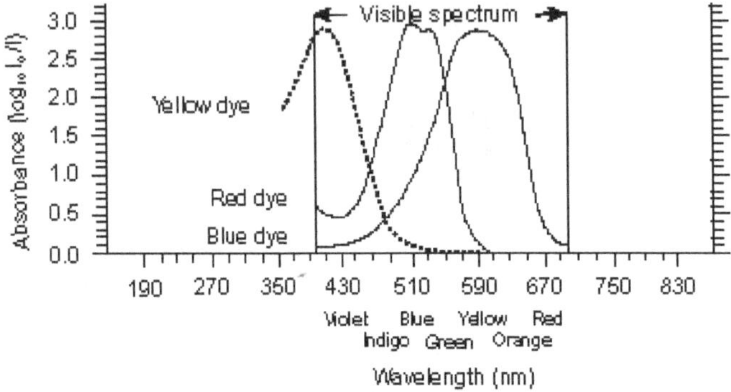

Figure 2.2   Absorption by coloured dyes

To understand why some compounds are coloured and others are not, and to determine the relationship of conjugation to colour, we must make accurate measurements of light absorption at different wavelengths in and near the visible part of the spectrum. Commercial optical spectrometers enable such experiments to be conducted with ease, and usually survey both the near ultraviolet and visible portions of the spectrum.

SINGLE AND DOUBLE BEAM SPECTROPHOTOMETERS

Instrumentation

In a single beam instrument, (Figure 2.3), the light transmitted through one sample is measured at a time. First, the solvent blank is placed in the cuvette and the instrument is adjusted to read zero transmittance with the light shutter closed, i.e., no light passes to the detector. Then, the shutter is opened and the instrument is adjusted to read 100% transmittance. The cuvette is removed and replaced by an identical one containing the sample. The absorbance is then read. The sequence of steps is maintained for every measurement. It means that both source output and detector sensitivity must remain constant, i.e., voltage supply to the source and detector should be stable. Such a procedure also compensates for reflection, scattering or absorption of light by the cuvette and the solvent.

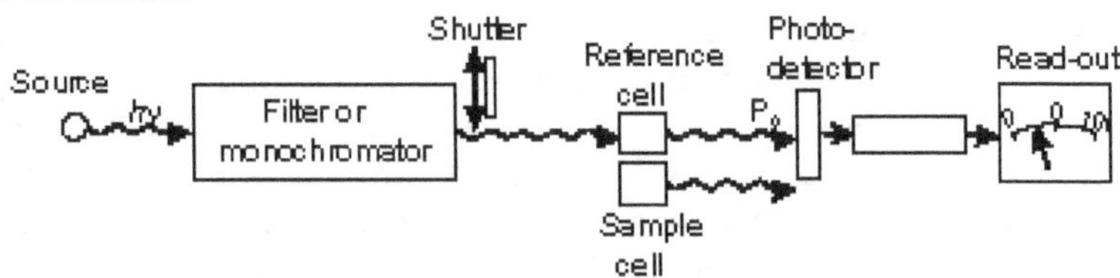

Figure 2.3   Single beam UV–visible spectrometer

The operation of the double beam instrument involves only two steps:

1. With the solvent blank in both beams, the recorder is adjusted to 100% transmittance.

2. With the solvent blank in reference beam and no radiation reaching the sample detector, the recorder is adjusted to full-scale deflection, i.e., zero transmittance.

Once the instrument is calibrated, the transmittance of sample can be measured at any wavelength. Signal from the sample beam is continually being referred to signal from the reference. Only the difference in the signal from the two beams is amplified and recorded. Thus, the output is independent of source and detector response with a change in wavelength.

If there is a drift in the signal, it affects the sample and reference beams equally. Thus, there is no analytical error in a double beam instrument. Slow variation of the average signal (not noise) with time is called **drift**. It may be due to deterioration of source or shifting of monochromator due to vibration or heating. The response of the detector varies with wavelength even if the light intensity is constant. The relationship between signal from detector and wavelength is called **response curve**. The signal starts from a low value, increases to a value that is steady over a wide range and then decreases again. This problem is overcome in a double beam instrument.

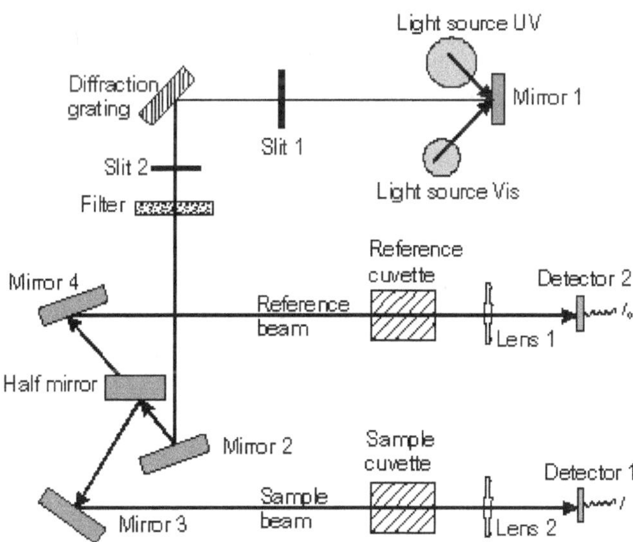

Figure 2.4   Components of a double beam spectrometer

The components of a typical double beam spectrometer are shown in Figure 2.4. The functioning of this instrument is relatively straightforward.

A beam of light from a visible and/or UV light source is separated into its component wavelengths by a prism or diffraction grating. Each monochromatic (single wavelength) beam in turn is split into two equal intensity beams by a half-mirrored device. One beam, the sample beam, which, passes through a small transparent container (cuvette) containing a solution of the compound being studied in a transparent solvent. The other beam, the reference beam, passes through an identical cuvette containing only the solvent. The intensities of these light beams are measured by electronic detectors and compared. The intensity of the reference beam, which should have suffered little or no light absorption, is defined as I_0. The intensity of the sample beam is defined as I. Over a short period of time, the spectrometer automatically scans all the component wavelengths in the manner described above. The UV region scanned is normally from 200 to 400 nm, and the visible portion is from 400 to 800 nm.

Radiation Sources

A good spectrophotometric source should have a stable, high-intensity output that covers a wide range of wavelengths. No single source is suitable for all of the spectral regions.

Sources are divided into thermal and electric discharge sources. Thermal radiation is the result of high temperature.

Deuterium discharge lamp This source uses an electrical discharge to dissociate deuterium molecules into atoms. The process is accompanied by the emission of continuous UV radiation in the range of about 160 to 380 nm. This is given by the following reaction.

$$D_2(g) \xrightarrow{\text{Electrical discharge}} D_2^*(g) \longrightarrow 2D(g) + h\nu$$

Internal energy of the excited deuterium molecules is converted into kinetic energy of the atoms and partly to radiant energy. Since the kinetic energies of the atoms are continuous (not quantized), the photon energies are also continuous. The discharge is carried out in a sealed tube containing data pressure around 1 to 5 torr between a heated filament and a metal electrode held at a potential difference of about 40 V. A quartz window must be built into the tube to allow the radiation to escape because glass absorbs strongly below 325 nm. Though the emission intensity is more with deuterium, it is very expensive. Hence, sometimes hydrogen discharge source is used.

Tungsten filament lamp The most common source of visible radiation is the ordinary tungsten filament lamp. It consists of a thin coiled tungsten wire sealed in an evacuated glass bulb. Electrical energy passing through the filament is converted to heat causing it to glow 'white hot'. The spectral distribution of an incandescent lamp is essentially the same as that of a black body radiator. The output intensity and wavelength distribution of a black body is independent of the material, depending only on the temperature. The filament temperature of W lamp is about 2900 K and the lamp output (325 to 3000 nm) covers a part of the UV, the entire visible and a sizable portion of the infrared spectrum. However, the intensity of the short wavelength (<350 nm) is small. To maintain a constant intensity, the electric current to the lamp should be controlled carefully.

The tungsten–halogen or quartz–halogen lamp is a W filament lamp that has been modified to operate at a higher temperature (about 3500 K) and, therefore, has a more intense output. The ordinary W lamp has a very short lifetime when operated at 3500 K because the filament vaporizes and eventually break. By adding a small amount of iodine to the evacuated tube, much of the vaporized W reacts with iodine to form tungsten iodide. This compound is relatively stable and remains in the bulb until it strikes the hot filament where it decomposes into W and iodine vapour. The process can be represented by the following reactions.

$$W(s) \longrightarrow W(g)$$

$$W(g) + I_2(g) \longrightarrow WI_x(g)$$

$$\underset{\text{hot filament}}{WI_x(g) + W(s) \longrightarrow 2W(s) + I_2(g)}$$

Ordinary glass becomes soft at higher operating temperature, and so the bulb is usually made of quartz, hence the name quartz–halogen lamp.

Mercury is used under high pressure in Hg discharge tubes. It is not suitable for continuous spectral studies because of the presence of sharp lines or bands superimposed on a continuous background.

Xenon discharge lamp operates with a low voltage DC source similar to that of the H lamp but at xenon pressures in the range of 10 to 30 atm. The intensity in the near UV is actually much greater than that of H lamp, but even greater intensity in the visible region may pose potential stray radiation problems.

Monochromators

The purpose of a monochromator is to disperse the radiation according to wavelength. Prisms and gratings are used extensively for this purpose. The most popular materials for making prisms are glass, quartz and fused silica. Of these, glass has the highest resolving power. It disperses light strongly over the visible region of the spectrum. However, it is not transparent to radiation with wavelength between 350 and 200 nm, because glass absorbs strongly and, therefore, cannot be used over this wavelength range.

Cuvettes

Virtually all UV spectra can be recorded in the solution phase and the samples are placed in cells or cuvettes. Cells may be made of plastic, glass or quartz. Quartz is transparent in all (200–700 nm) ranges and is normally used in UV region. Plastic and glass are only suitable for visible spectra. A typical cell (commonly called a cuvette) is shown in Figure 2.5.

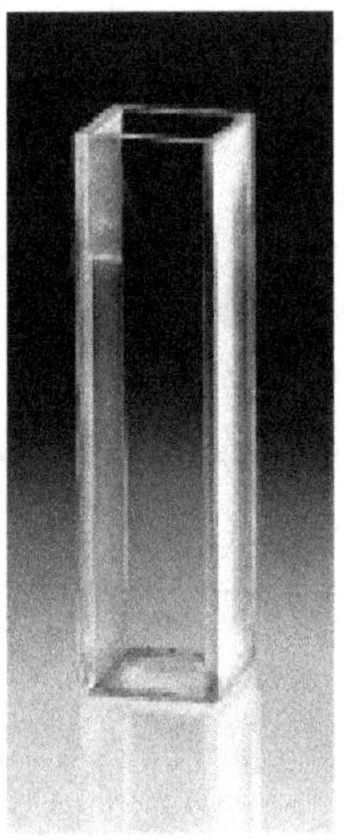

Figure 2.5   A typical cell (cuvette)

Detectors

The four common types of detectors are:

1. Barrier layer cell or photovoltaic cell
2. Phototube or photocell or photo-emissive tubes
3. Photomultiplier tubes (PMT)
4. Semiconductor devices:
   (a) Light dependent resistor (LDR) and
   (b) Linear photodiode array (LPDA)

Barrier layer cell   Silver is coated on a semiconductor, such as Se, that is joined to a strong metal base such as iron, by vaporization method Figure 2.6. Any radiation falling on the surface generates electrons at the Se–Ag surface. The barrier existing between Se and iron prevents electrons from flowing into the iron. The electrons are, therefore, collected by silver. The accumulation of electrons creates an electric voltage difference between silver surface and the base of the cell. The voltage produced is a measure of the intensity of radiation falling on it.

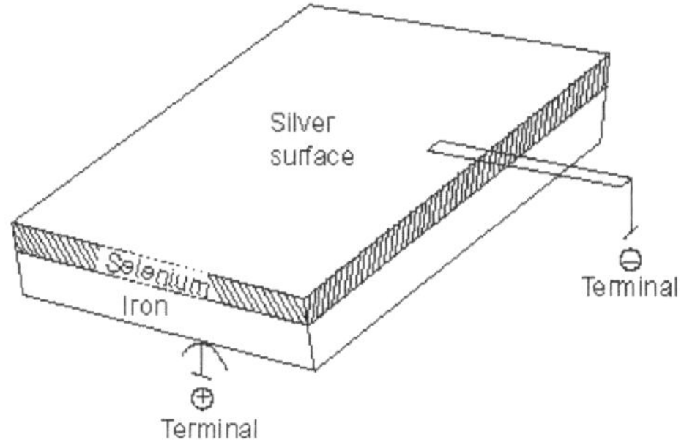

Figure 2.6 Barrier layer cell

Phototube A phototube contains a photocathode and anode but not dynodes. It consists of a large semi-cylindrical cathode and a small concentric anode enclosed in an evacuated glass envelope (Figure 2.7). If the phototube is to be used in ultraviolet region, a window of quartz is mounted on the wall of glass tube to allow radiation to pass into the tube. The cathode is coated with photo-emissive substances such as Co_3Sb, K_2CsSb and Na_2KSb mixed with traces of Cs.

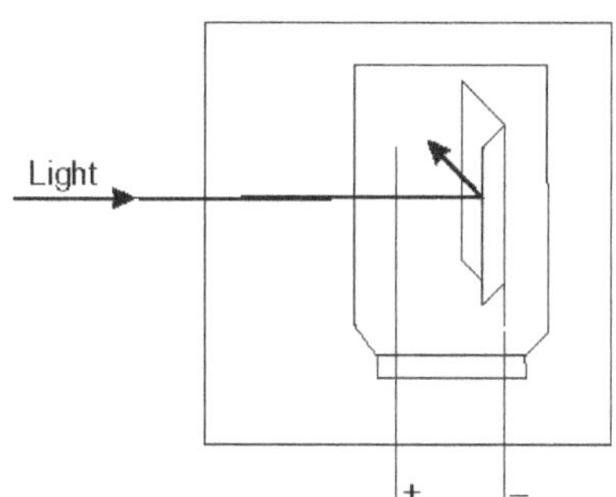

Figure 2.7 Phototube

A voltage of 90 V is maintained between cathode and anode by an external power supply. The measured potential is directly proportional to the current flowing through the circuit. In the absence of radiation, no current flows through the circuit. As radiation strikes the cathode, electrons are emitted from cathode and accelerated to the anode. This causes a current flow in the circuit. The rate of electron emission from the cathode is proportional to the intensity of radiation. Therefore, measured voltage is proportional to the intensity of radiation.

Photomultipliers These are sensitive, fast-responding vacuum phototubes in which an amplification of several millionfold is achieved within one tube by the emission of secondary electrons. The radiation striking the photocathode causes ejection of primary electrons. These electrons are accelerated by a positive potential to a second sensitive surface. Here, the number of electrons is increased by a factor of about ten (Figure 2.8 and 2.9). This process is repeated many times (10 target electrodes or dynodes). Each dynode is made 75–100 V more positive than the preceding dynode. The cathode is coated with Ca_3Sb, K_2CsSb and Na_2KSb. Dynodes are coated with BeO, CsSbor GaP.

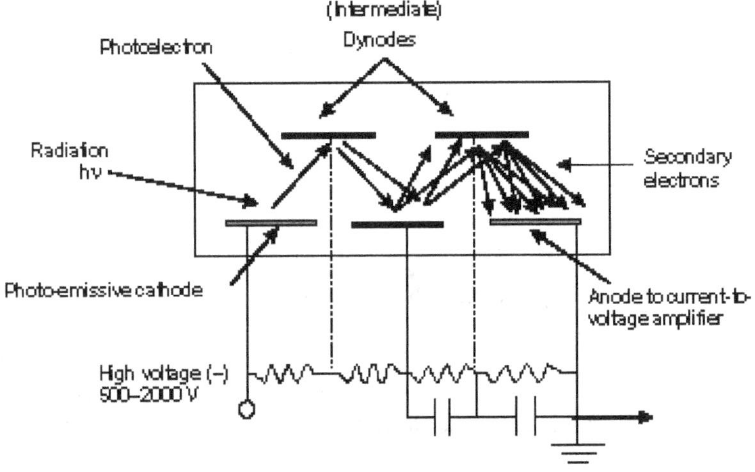

Figure 2.8 Schematic representation of flow of electrons in PMT

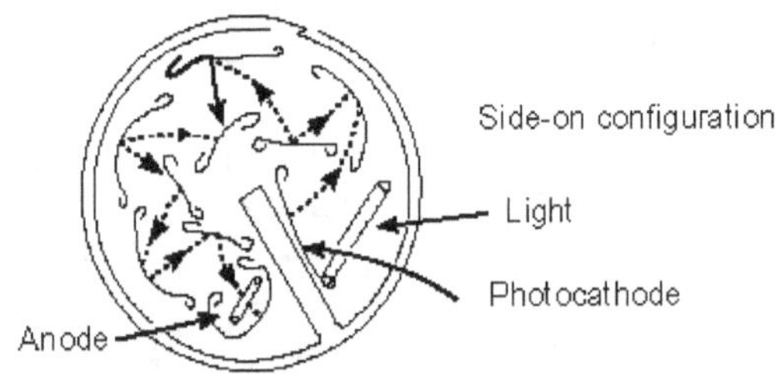

Figure 2.9 Photomultiplier tube

Photodiode and LPDA   In a photodiode, *p–n* junction is constructed. It consists of high-resistivity intrinsic silicon material. Very shallow *p* and *n* diffusions are made at the top and bottom surfaces. The top surface is covered with silicon dioxide layer. The diffused *p* regions determine the junction and optically active areas. A photon must reach the active area to produce current flow in the external circuit. The *p–n* junction is reverse biased so that no current flows. When incident ray falls on the diode, electrons are promoted to the conduction band and this generates a current proportional to incident radiant power.

The linear photodiode array (LPDA) (Figure 2.10) enables simultaneous measurement of light intensity and has absorption at many wavelengths. The radiation from source passes through the sample to the grating. The radiation is separated by wavelengths and directed to a diode array system. The diode array consists of a number of semiconductors embedded in a single crystal. Individual diodes are called elements or channels. If radiation falls on the system, the photon will displace an electron in the n type silicon. These electrons go to the nearest p type semiconductor under the influence of reverse biased p–n junction. The current produced depends on the intensity of light falling on the system. By measuring the charges on each individual element, it is possible to get a measurement of light intensity versus wavelength of the whole range.

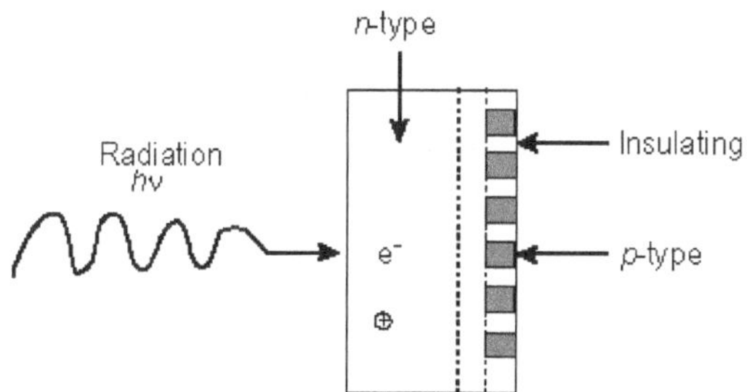

Figure 2.10   Schematic representation of a LPDA

Read-out   With simpler instruments, direct current (DC) signals are produced, which are amplified and displayed on meters, recorders or digital voltmeters. However, DC amplifiers are subject to drift and offset errors. It is, therefore, often desirable to modulate the signal and transform it into alternating current of frequency high enough to avoid drift and noise problems. After amplification by an AC amplifier, the signal is converted back into DC by a demodulator because the read-out devices require DC signals. In an x-y recorder, an absorption curve relating wavelength (abscissa) and absorbance (ordinate) is plotted.

Good Operating Techniques

Spectrophotometry is most accurate at intermediate levels of absorbance (A = 0.4 to 0.9). If too little light gets through the sample (high absorbance), the intensity is hard to measure. If too much light gets through the sample (low absorbance), it is hard to distinguish the transmittance of the sample from that of the reference. Therefore, the concentration of the sample should be adjusted such that the absorbance is in the range 0.4 to 0.9.

For spectrophotometric analysis, measurements are made at a wavelength corresponding to a peak in the absorbance spectrum. This wavelength gives the greatest sensitivity (maximum response) for a given concentration of analyte. Errors due to drift and finite bandwidth of wavelengths selected by the monochromator are minimized because the spectrum varies least with wavelength at the absorbance maximum.

Cuvettes should be handled with a tissue to avoid fingerprints and must be kept scrupulously clean to avoid surface contamination, which may lead to scattering.

Theory of Electronic Spectroscopy

The energies of visible and ultraviolet regions are sufficient to promote or excite a molecular electron to a higher energy orbital. Absorption spectroscopy carried out in this region is sometimes called **electronic spectroscopy**. A diagram showing various possible kinds of electronic excitation in organic molecules is shown in Figure 2.1 and the $\pi \rightarrow \pi^*$ transition is shown in Figure 2.12. Of the six transitions outlined, only the two ones with lowest energy are achieved by the energies available in the 200 to 800 nm spectrum. As a rule, energetically favoured electron promotion will be from the **highest occupied molecular orbital (HOMO)** to the **lowest unoccupied molecular orbital (LUMO),** and the resulting species is called an **excited state**.

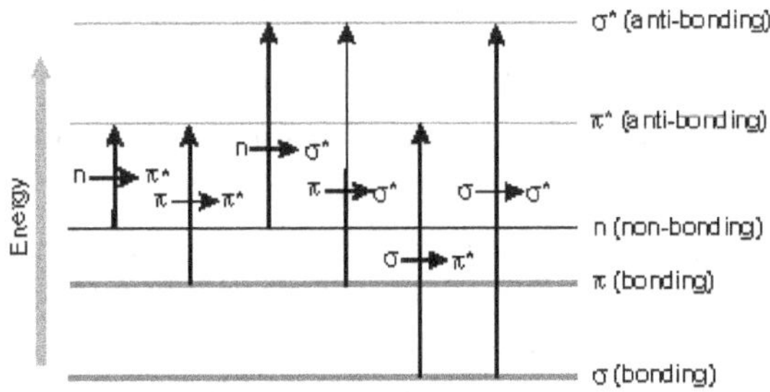

Figure 2.11   Various electronic excitations in organic molecules

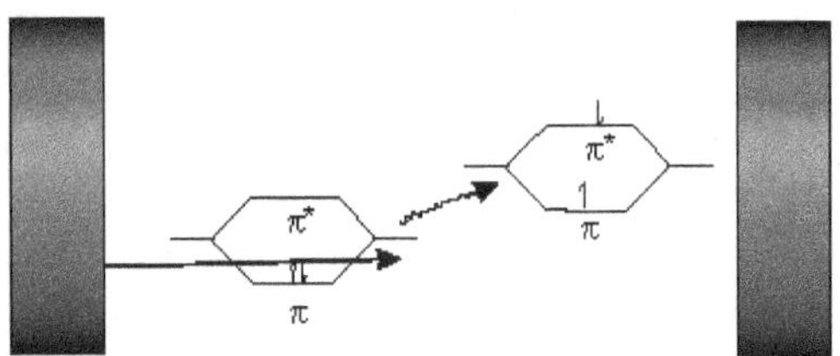

Figure 2.12 $\pi \rightarrow \pi^*$ transition

When sample molecules are exposed to light having an energy that matches a possible electronic transition within the molecule, some of the light energy will be absorbed as the electron is promoted to a higher energy orbital. An optical spectrometer records the wavelengths at which absorption occurs, together with the degree of absorption at each wavelength. The resulting spectrum is presented as a graph of absorbance (A) versus wavelength, as in the isoprene spectrum shown in Figure 2.13. Since isoprene is colourless, it does not absorb in the visible part of the spectrum and this region is not displayed on the graph.

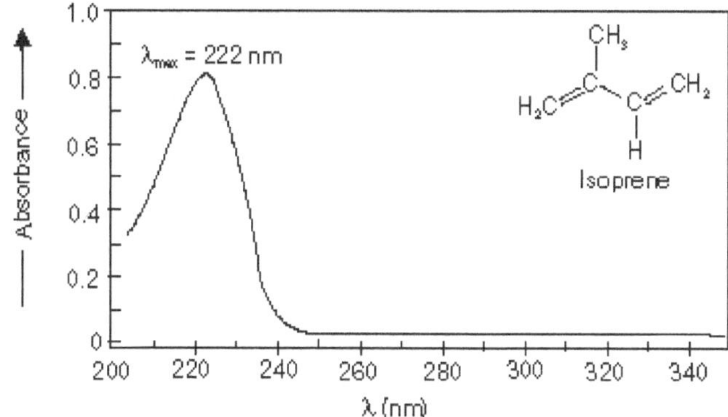

Figure 2.13   UV spectrum of isoprene

For most spectra, the solution obeys Beer's law. This law states that the light absorbed is proportional to the number of absorbing molecules, i.e., concentration of absorbing molecules. This is only true for dilute solutions.

$$\log \frac{I_o}{I} \propto c$$

A second law—Lambert's law—tells us that the fraction of radiation absorbed is independent of the intensity of the radiation and a linear relationship exists between absorbance and cell path length.

$$\log \frac{I_o}{I} \propto l$$

Combining these two laws gives the Beer–Lambert law:

$$\log I_o / I = \varepsilon l c$$

where,

I_o = the intensity of the incident radiation

I = the intensity of the transmitted ε radiation

  = the molar absorption coefficient

l = the path length of the absorbing solution (cm)

c = the concentration of the absorbing species in mol dm^{-3}.

Absorbance usually ranges from 0 (no absorption) to 2 (99% absorption), and is precisely defined in context with spectrometer operation. Because the absorbance of a sample will be proportional to the number of absorbing molecules in the spectrometer light beam (e.g. their molar concentration in the sample tube), it is necessary to correct the absorbance value for this and other operational factors if the spectra of different compounds are to be compared in a meaningful way. The corrected absorption value is called 'molar absorptivity', and is particularly useful when comparing the spectra of different compounds and determining the relative strength of light- absorbing functions (chromophores). **Molar absorptivity** (e) is defined as the ratio of absorbance to the sample concentration and the path length of light through the sample.

$$\varepsilon = A/cl$$

where,

  A = absorbance

  c = sample concentration in mol/l

  l = length of light path through the sample in cm.

If the isoprene spectrum was obtained from a dilute hexane solution ($c = 4 \times 10^{-5}$ moles per litre) in a 1-cm sample cuvette, a simple calculation using the above formula indicates a molar absorptivity of 20,000 at maximum absorption wavelength. Indeed, the entire vertical absorbance scale may be changed to a molar absorptivity scale once this information about the sample is available.

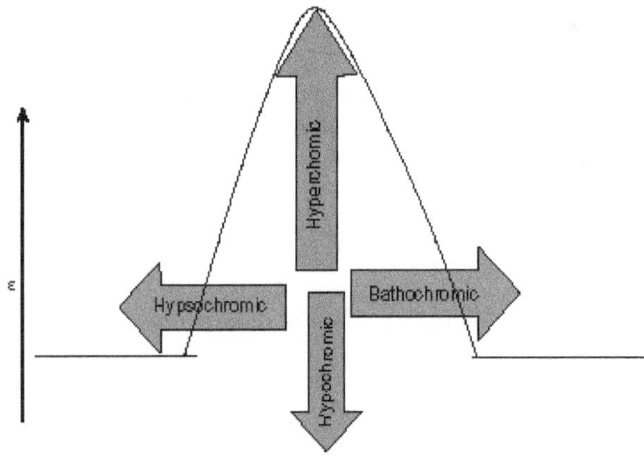

Figure 2.14   Shifts in UV–visible spectrum

The absorption maximum can be altered by solvents or substituents.

The Chromophore Concept

All those compounds, which absorb in the 400 to 800 nm range appear coloured. Exact colour depends on the wavelength of light absorbed. This in turn depends on the presence of specific groups called chromophores. Orginally a chromophore was considered as any system, which is responsible for imparting colour to the compound. Now the term chromophore is used in a broader sense. It is defined as any isolated covalently bonded group that shows a characteristic absorption in the UV or the visible region.

There are two types of chromophores.

1. *Independent chromophore* When a single chromophore is sufficient to impart colour, e.g. $-NO_2$, $-NO$, o and p quinonoid groups.

2. *Dependent chromophore* When more than one chromophore is required to impart colour, e.g. CH_3COCH_3 (colourless), $CH_3COCOCH_3$ (yellow) $CH_3COCOCOCH_3$ (orange).

Auxochromes The saturated groups with nonbonding electrons, when attached to chromophores alter the wavelength and intensity of absorption. These are called auxochromes, e.g. $-OH$, $-NH_2$, $-Cl$.

Absorption and intensity shifts

1. **Bathochromic shift** It is an effect by virtue of which the absorption maximum is shifted towards longer wavelength due to the presence of an auxochrome or by the change of the solvent. It is also called red shift.

2. **Hypsochromic shift** It is an effect by virtue of which the absorption maximum is shifted towards shorter wavelength. It is also called blue shift.

3. **Hyperchromic shift** It is an effect due to which the intensity of absorption maximum increases. The introduction of an auxochrome usually increases the intensity of absorption.

4. **Hypochromic shift** It is defined as an effect due to which the intensity of absorption maximum decreases.

General aspects of UV studies

1. UV peaks or bands are classified into three groups— high or strong (s) intensity peaks ($\varepsilon > 104$), moderate or medium (m) intensity peaks ($\varepsilon \approx 103\text{–}104$), and weak (w) or low intensity peaks ($\varepsilon < 103$).

2. The energy content of UV radiation is high. A molecule after absorption of UV radiation will be in an excited state, called a singlet excited state (ES). The ES molecule tends to lose its energy in several possible ways and return to the original ground state (GS). The life time of ES is short ($\approx 10^{-11}$ s). The ES molecule may

   * crossover to another ES called the triplet state which is more stable (lifetime $\approx 10^{-6}$ s).

   * lose the excess energy in the form of vibration and rotation, and return to GS.

   * emit energy in the form of radiation.

   * undergo exothermic chemical reactions.

3. Changes in temperature do not affect a UV spectrum directly. However, vibrational energy may be affected by temperature, and UV bands will be sharper at lower temperatures. Changes in temperature may affect hydrogen bonding, association and disassociation of the compound and, thus, may affect the UV spectrum indirectly.

4. Experimental errors in the measurement of intensity are minimum in the middle range of absorption of a UV peak, i.e., when the transmittance is in the 15–65% range.

5. Accurate determination of intensity may be complicated by background absorption, i.e., in the absence of a compound, the spectrum does not coincide with the base line (zero absorption). In several cases, the base line of the spectrum may be parallel to the x-axis.

6. The UV peaks are not sharply defined signals but smoothed-out curves that extend over several nanometers. Frequently, two or more peaks occur close to each other and coalesce.

Selection Rules

Even when the energy of a photon matches the energy difference between two levels, the absorption of radiation may not be observed or may be observed with low intensity. The reason for this is that certain requirements called selection rules must be satisfied for a transition to occur.

The most important requirement is that an electron must be promoted without a change in spin orientation, i.e., $\Delta S = 0$. Transition from singlet to triplet state is multiplicity forbidden.

A transition is possible only if the symmetry of the molecular orbitals (MOs) change from g $\rightarrow$ u or u $\rightarrow$ g and is not possible if the change is g $\rightarrow$ g or u $\rightarrow$ u. This rule is applicable to molecules with centre of symmetry and the forbidden transitions are referred to as Laporte forbidden.

However, many of the forbidden transitions are found to occur. Some of the causes are

1. Symmetry may change during vibrations.

2. The singlet to triplet transitions of the excited state may take place so rapidly that the forbidden GS singlet to ES triplet may appear to occur. That is, two transitions GS $\rightarrow$ SES (singlet) $\rightarrow$ ES (triplet) may occur very rapidly.

3. Sometimes transitions are found to take place by absorption of energy not from radiation but from the excited state of another chromophore.

Empirical Rules for Absorption Wavelengths of Conjugated Systems

Woodward–Fieser Rules for Dienes

For dienes and trienes, the position of the most intense band can be correlated with the substituents present. Based on the polarity of the substituents, their contribution to the shift changes. The absorption moves progressively towards the longer wavelength as the number of conjugated double bond increases. Based on these factors, the expected absorption maximum can be empirically worked out. Woodward and Fieser worked out such empirical rules and found successful in predicting the λ_{max} of the compounds.

	Homoannular (cisoid)	**Heteroannular (transoid)**
Parent	$\lambda = 253$ nm	$\lambda = 214$ nm $= 217$ (acyclic)
Increments for		
Double bond extending conjugation	30	30
Alkyl substituent or ring residue	5	5
Exocyclic double bond	5	5
Polar groupings		
$-OC(O)CH_3$	0	0
$-OR$	6	6
$-Cl, -Br$	5	5
$-NR_2$	60	60
$-SR$	30	30

λ_{max} (calculated) = Base (214 or 217 or 253) + Substituent contributions

The following examples illustrate the effectiveness of the rules.

1. 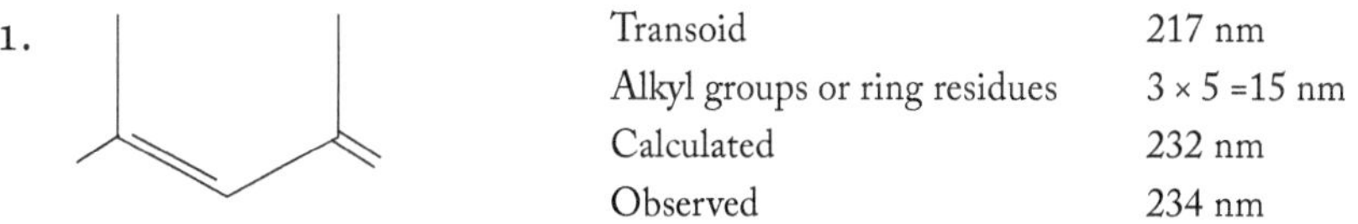

Transoid	217 nm
Alkyl groups or ring residues	$3 \times 5 = 15$ nm
Calculated	232 nm
Observed	234 nm

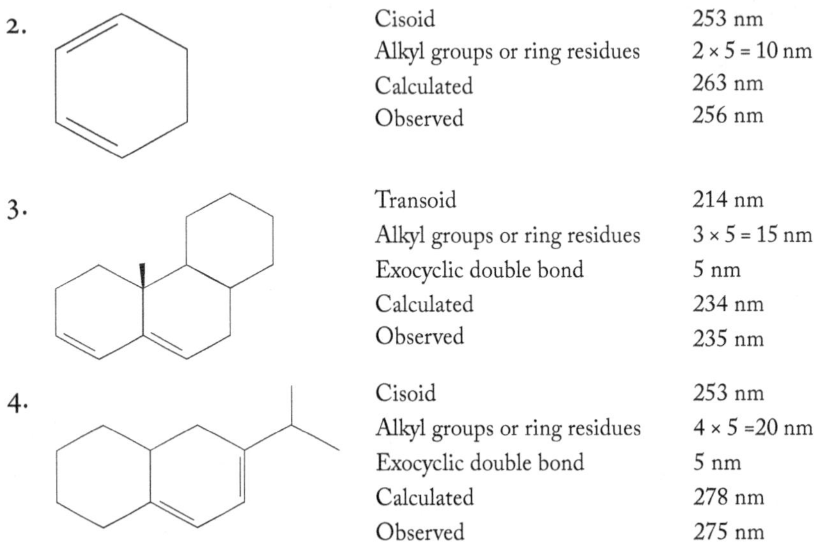

2.

Cisoid	253 nm
Alkyl groups or ring residues	$2 \times 5 = 10$ nm
Calculated	263 nm
Observed	256 nm

3.

Transoid	214 nm
Alkyl groups or ring residues	$3 \times 5 = 15$ nm
Exocyclic double bond	5 nm
Calculated	234 nm
Observed	235 nm

4.

Cisoid	253 nm
Alkyl groups or ring residues	$4 \times 5 = 20$ nm
Exocyclic double bond	5 nm
Calculated	278 nm
Observed	275 nm

Applications to organic systems Some examples that illustrate the above rules are as follow:

UV Data Sheet

Absorption maxima (λ_{max}) are in nanometers, molar absorptivities (ε) are in parentheses.

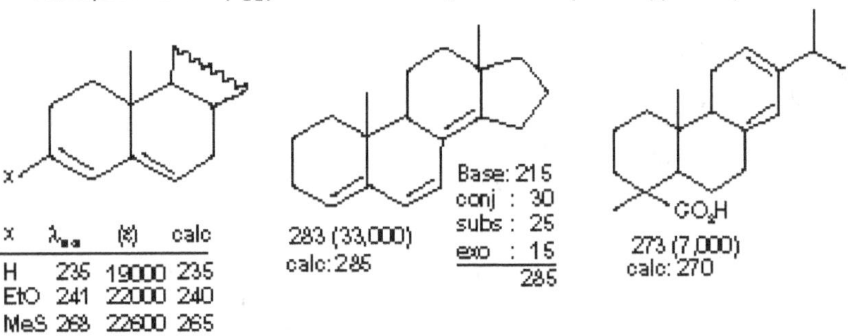

(*Contd.*)

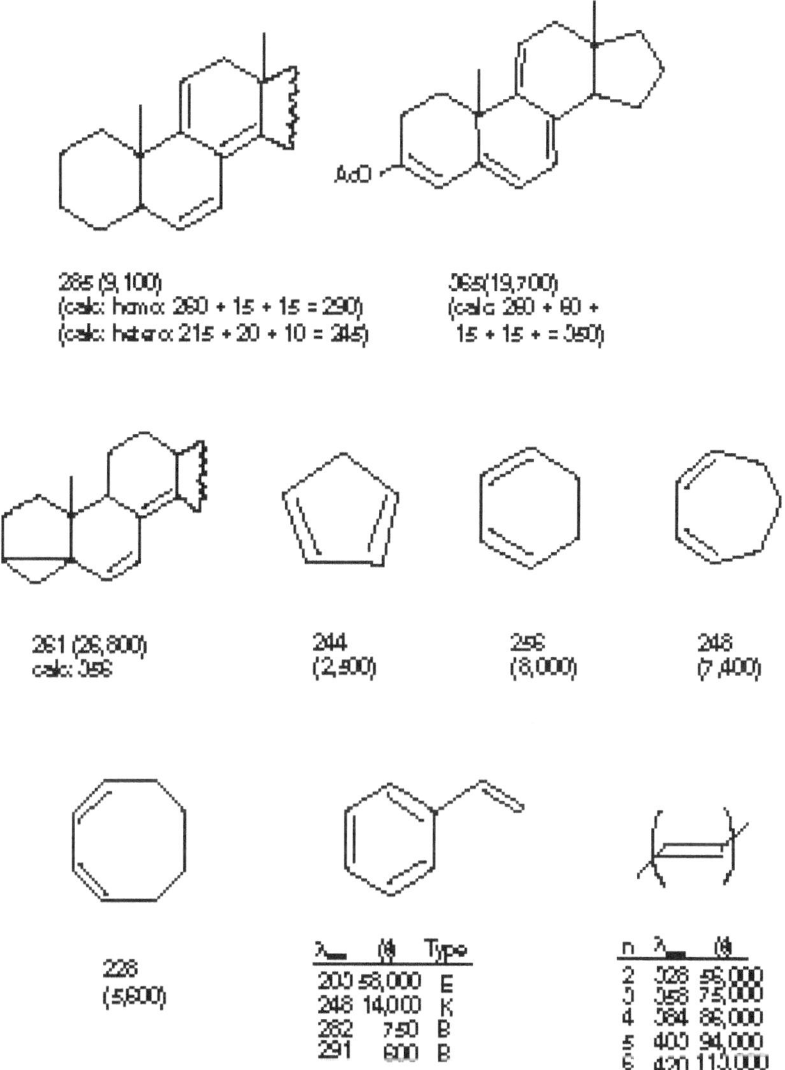

285 (9,100)
(calc: homo: 260 + 15 + 15 = 290)
(calc: hetero: 215 + 20 + 10 = 245)

365(19,700)
(calc 280 + 60 +
15 + 15 + = 350)

261 (26,800)
calc: 358

244
(2,500)

256
(8,000)

248
(7,400)

228
(5,600)

λ_{max}	(ε)	Type
200	58,000	E
248	14,000	K
282	750	B
291	600	B

n	λ_{max}	(ε)
2	328	56,000
3	358	75,000
4	384	86,000
5	408	94,000
6	420	113,000

As in the case of dienes, the α,β unsaturated carbonyl compounds can als $\pi \to \pi^*$ give rise to transition in the UV region apart from a weak $\pi \to \pi^*$ bond. Correlations that apply to enones were compiled by Woodward. The important differences between dienes and enones towards the use of these set of rules are a) the increment values towards the presence of different substituents differ depending on the position in the case of enones b) the effect of solvent is to be taken into account here but solvent effect is almost negligible in the case of dienes.

Woodward's Rules for Conjugated Carbonyl Compounds

Base values:

X = R

Six-membered ring or acyclic parent enone	$\lambda = 215$ nm
Five-membered ring parent enone	$\lambda = 202$ nm
X = H	$\lambda = 208$ nm
X = OH, OR	$\lambda = 195$ nm

Increments for:

Double bond extending conjugation		30
Exocyclic double bond		5
Endocyclic double bond in a 5- or 7-membered ring for X = OH, OR		5
Homocyclic diene component		39
Alkyl substituent or ring residue	α	10
	β	12
	γ or higher	18

Polar groupings:

—OH	α	35
	β	30
	δ	50
—OC(O)CH$_3$	$\alpha, \beta, \gamma, \delta$	6
—OCH$_3$	α	35
	β	30
	γ	17
	δ	31
—Cl	α	15
	β, γ, δ	12
—Br	β	30
	α, γ, δ	25
—NR$_2$	β	95

Solvent shifts for various solvents:

Solvent	λ_{max} shift (nm)
Water	+ 8
Chloroform	−1
Ether	−7
Cyclohexane	−11
Dioxane	−5
Hexane	−11

λ_{max} (calculated) = Base + Substituent contributions and corrections

The examples given below illustrate the utility of these rules.

1.

Acyclic enone	215 nm
α-Alkyl groups or ring residues	10 nm
β-Alkyl groups or ring residues	2 × 12 = 24 nm
Calculated	249 nm
Observed	249 nm

2.

Five-membered ring parent enone	202 nm
β-Alkyl groups or ring residues	2 × 12 = 24 nm
Exocyclic double bond	5 nm
Calculated	231 nm
Observed	226 nm

3.

Six-membered ring or alicyclic parent enone	215 nm
	30 nm
Extended conjugation	39 nm
Homocyclic diene component	18 nm
δ-Alkyl groups or ring residues	302 nm
Calculated	300 nm
Observed	

4.

Five-membered ring parent enone	202 nm
α-Br	25 nm
β-Alkyl groups or ring residues	2 × 12 = 24 nm
Exocyclic double bond	5 nm
Calculated	256 nm
Observed	251 nm

5.

Carboxylic acid	195 nm
α-Alkyl groups or ring residues	10 nm
β-Alkyl groups or ring residues	12 nm
Calculated	217 nm
Observed	217 nm

6.

Ester	195 nm
α-Alkyl groups or ring residues	10 nm
β-Alkyl groups or ring residues	12 nm
Endocyclic double bond in 7-membered ring	5 nm
Calculated	222 nm
Observed	222 nm

7.

Aldehyde	208 nm
α-Alkyl groups or ring residues	10 nm
β-Alkyl groups or ring residues	2 × 12 = 24 nm
Calculated	242 nm
Observed	242 nm

8.

Aldehyde	208 nm
Extended conjugation	30 nm
Homodiene component	39 nm
α-Alkyl groups or ring residues	10 nm
δ-Alkyl groups or ring residues	18 nm
Calculated	304 nm
Observed	302 nm

Applications to organic systems   Some examples that illustrate the above rules ae as follows.

UV Data Sheet

Absorption maxima (λ_{max}) are in nanometres, molar absorptivities (ε) are in parentheses.

233 (13,000)
calc: 237

245 (6,800)
calc: 249

243 (1,400)
calc: 249

246 (12,300)
calc: 249

245 (5,800)
calc: 242

281 (14,000)
calc: 283

245 (15,000)
calc: 244 or 237

(Contd.)

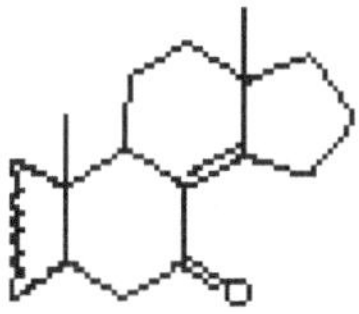

259 (9,500)
calc: 259

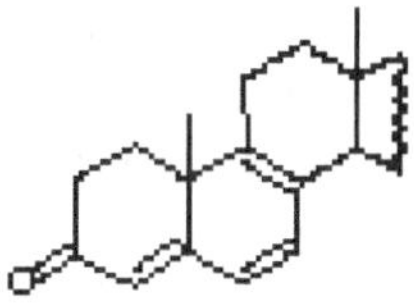

388 (18,000)
calc: 381

Base:	215
conj:	60
subs:	68
exo	5
6-ring:	35
	381

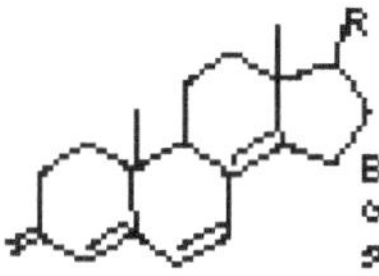

348 (40,000)
calc: 358

Base:	215
conj:	60
subs:	68
exo	15
	358

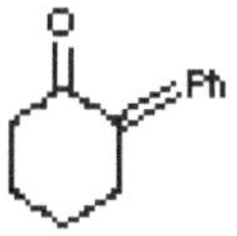

294 (17,000)
calc: 290
215 + 10 + 5 + 60 = 290

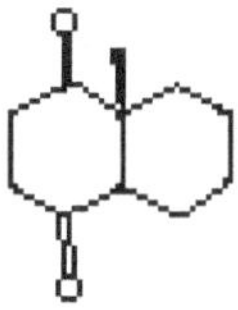

225
calc: 225

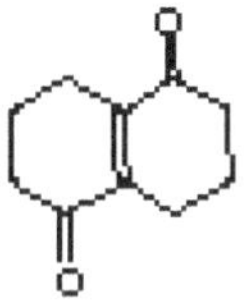

275
calc: 249

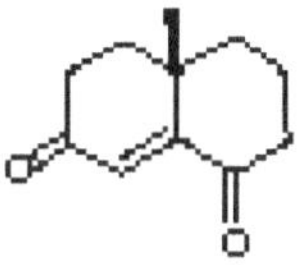

252
calc: 244

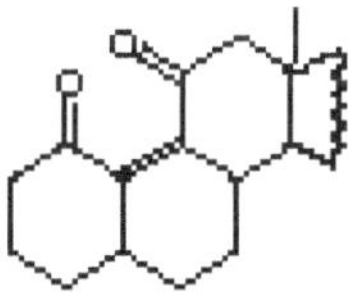

280
calc: 259

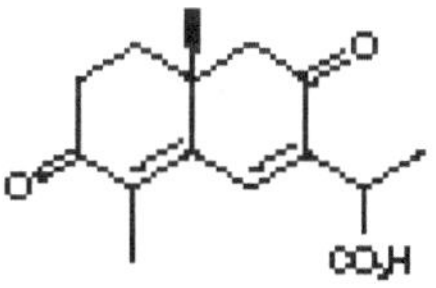

311 (20,000)
calc: 309 or 314

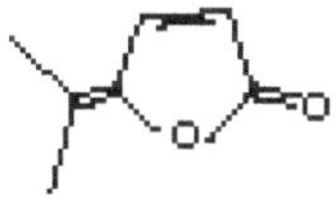

270 (23,000)
calc: 273 or 278

Electronic spectra of complex compounds

Russell–Saunder's Coupling Scheme

Electronic spectra of coordination compounds can be explained based on microstate, spectroscopic state, J state and group theoretical term.

Taking microstate, let us consider a simple example namely free carbon atom, C with electronic configuration $1s^2, 2s^2, 2p^2$.

The two electrons in the 1s-orbital can be arranged in only one way; the orbital is completely filled. Similarly, the two electrons in the 2s-orbital can be arranged in only one way. When 2p-orbitals are considered, the answer is to be found for the number of ways the two electrons can be arranged with the available three p-orbitals, namely $2p_x$, $2p_y$ and $2p_z$. The answer can be found mathematically, thus:

$$\frac{6 \times 5}{2!} = \frac{6 \times 5}{2 \times 1} = 15$$

6, because six places are available for the first electron;

5, because five places are available for the second electron;

2!, because two electrons are considered.

Fifteen arrangements are shown (Figure 2.15) and each arrangement is called a microstate. Hence, free carbon atom has got 15 microstates. Consider the first microstate:

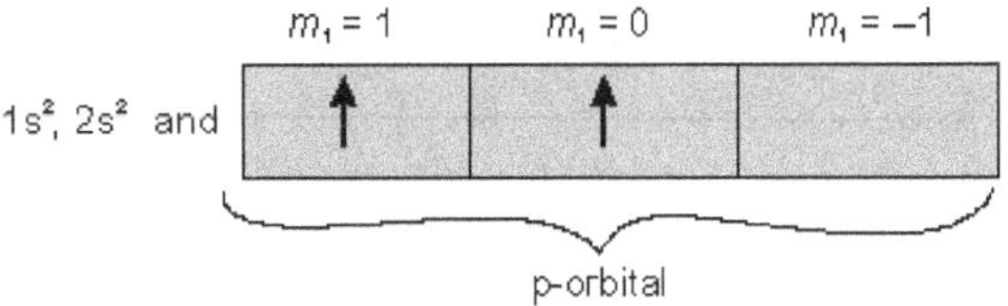

where,

m_l = 1 stands for the first p-orbital

m_l = 0 stands for the second p-orbital and

m_l = −1 stands for the third p-orbital.

There are six electrons and there will be repulsion among electrons; the total inter-electronic repulsion value may be calculated. Similarly, inter-electronic repulsion value may be calculated for each of the remaining microstates.

Among the 15 microstates, some may have the same value for the total inter-electronic repulsion; they may be grouped together in a family and given the name **spectoscopic state.**

Name of Family	Total Inter-Electronic Repulsion	Number of Microstates in a Family
Spectroscopic State 1	Say, X	9
Spectroscopic State 2	Say, Y	5
Spectroscopic State 3	Say, Z	1

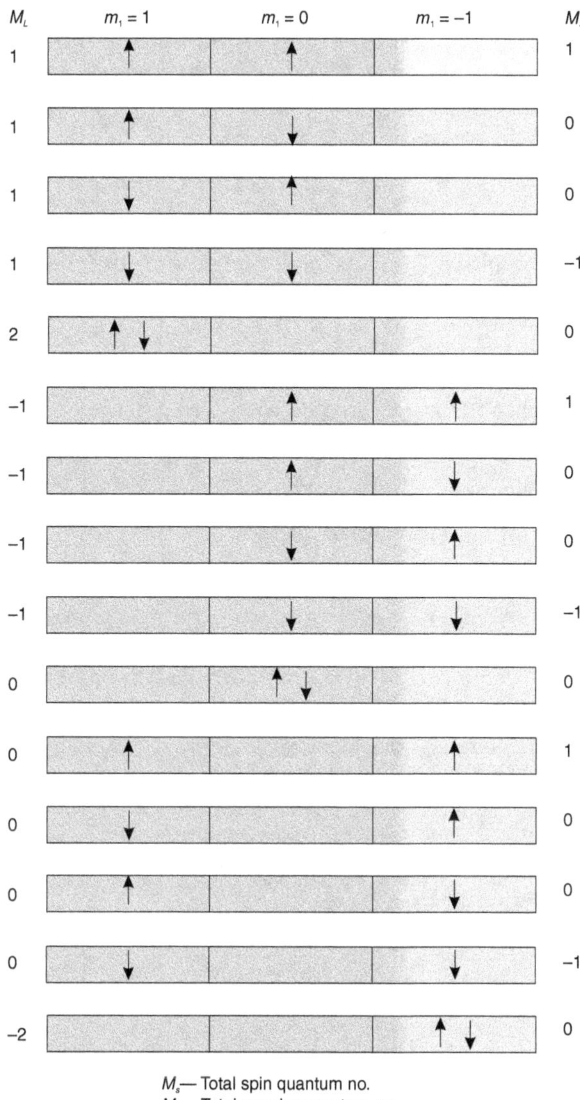

Figure 2.15 The 15 microstates of a free carbon atom

This is explained as follows:

Consider the two p-electrons in free carbon atom. Each electron is having a spin angular momentum and an orbital angular momentum. Let s_1 and s_2 represent the spin angular momentum of p-electron 1 and p-electron 2, respectively; l_1 and l_2 represent the orbital angular momentum of p-electron1 and p-electron 2, respectively.

p-electron 1   p-electron 2

s_1 s_2

l_1 l_2

s_1 and s_2 are coupled to give total spin angular momentum, (S), i.e.,

$$\Sigma_i s_i = S$$

l_1 and l_2 are coupled to give total orbital angular momentum (L), i.e.,

$$\Sigma_i l_i = L$$

$l = 2$ corresponds to d.

$l = 2$ represents the five d-orbitals with corresponding m_l values 2, 1, 0, −1 and −2.

Suppose m_l values are 1, 0 and −1; the l value is 1.

What is said for *small l* applies to *capital L*.

If we assign M_L value and M_S value for each microstate of the free carbon atom (Figure 2.15) then,

Maximum M_L is 2. $M_L = 2$ should have come from $L = 2$.

If $L = 2$, possible M_L values are 2, 1, 0, −1, −2.

Look for maximum M_S value along with maximum M_L value, which is 0.

$M_S = 0$ should have come from $S = 0$.

If $S = 0$, possible M_S value is 0 only.

M_L	M_S	Microstates
2	0	1
1	0	2
0	0	3
-1	0	4
-2	0	5

$L = 2$, therefore D state.

If $S = 0$, the value of $2S + 1 = 1$.

∴ Spin degeneracy or spin multiplicity is 1.

∴ Spectroscopic state is ^1D (called as *singlet* 'dee').

In ^1D, the superscript refers to spin degeneracy.

Orbital degeneracy is 5 (i.e., $2L + 1$); hence, spin degeneracy × orbital degeneracy is $1 \times 5 = 5$, which is the number of microstates in the spectroscopic state. Total inter-electronic repulsion among six electrons is the same in all the five microstates in ^1D, and that is why they are grouped together in a family called *singlet 'dee'*.

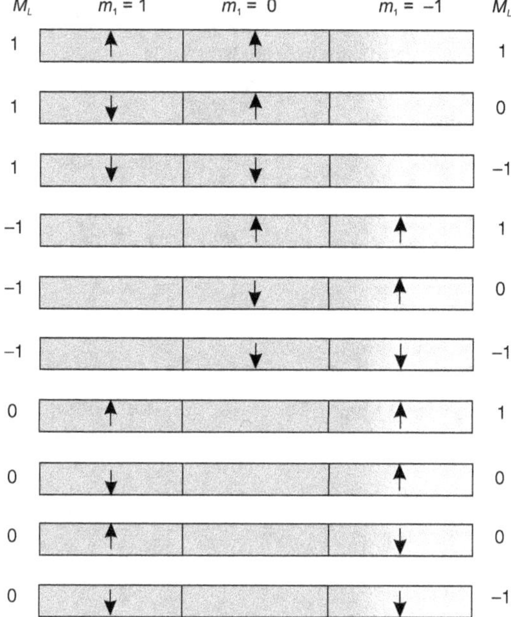

Figure 2.16   The 10 microstates of free carbon atom

Now, remove the above five microstates from the total of 15 microstates and look at the remaining 10 microstates. The remaining 10 microstates are shown in Figure 2.16. Follow the same procedure as above to derive the second spectroscopic state ^3P.

^3P means L = 1, S = 1.

$(2L + 1) = 3$ is the orbital degeneracy of the spectroscopic state.

$(2S + 1) = 3$ is the spin degeneracy of the spectroscopic state.

$(2L + 1) \times (2S+1) = 3 \times 3 = 9$ is the total number of microstates of the spectroscopic state.

Remove the nine microstates of ^3P from the above 10 microstates and then find the remaining one microstate, which is:

This microstate is the only one belonging to the spectroscopic state ^1S. Thus, free carbon atom contains the spectroscopic states ^1D, ^3P and ^1S, and the state with maximum spin degeneracy, namely ^3P, will be the ground state.

J state can be explained by choosing lanthanide ions as examples as they behave like free ions, whatever be the environment (F electrons are shielded effectively by outer electrons from the environment). The spectroscopic state of a free atom/ion possesses L (total orbital angular momentum) and S (total spin angular momentum), and the coupling between L and S gives J; it is more effective in the case of lanthanide ions as l, the spin orbit coupling constant, is large. Consider the tri-positive cerium ion Ce^{3+}, which has got only one spectroscopic state namely ^2F. For ^2F, L =3; S = 1/2 and J can have vlues from $|L+S|$ to $|L-S|$ so that successive values differ by unity. Thus nthis case,

$$|L+S| \qquad |L-S| \qquad \left(3+\frac{1}{2}\right)\left(3-\frac{1}{2}\right)$$

J = 7/2 .... 5/2 - only two values are possible.

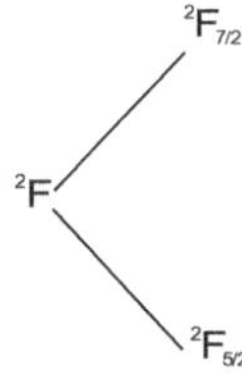

Ce³⁺ ion Ground J state is $^2F_{5/2}$ as it is an f^1 system which is less than half-filled, namely f^7 and hence the lowest J = 5/2, is the ground J state.

Consider another example, Yb^{3+} ion

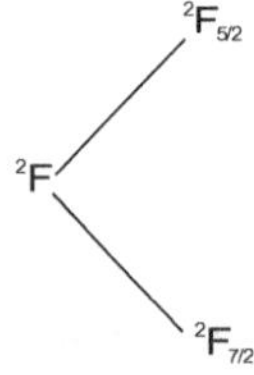

Yb³⁺ ion Ground J state is $^2F_{7/2}$ as it is an f^{13} system which is more than half-filled, namely f^7 and hence the highest J = 7/2, is the ground J state.

Crystal Field Spectra of 3d¹ System

In an octahedral environment, the number of microstates for 'free' (absence of any crystal field) d^1 ion is 10.

Consider a d^1 ion in an octahedral environment. The ground state electronic configuration will be $(t_{2g})^1$ and the ground term is described by the group theoretical term $^2T_{2g}$. Similarly, the excited state electronic configuration will be $(e_g)^1$ and the excited term is described by the group theoretical term ^2E_g. One absorption band is predicted corresponding to the electronic transition $(t_{2g})^1 \rightarrow (e_g)^1$ and depicte as $^2E_g{}^2 \leftarrow T_{2g}$.

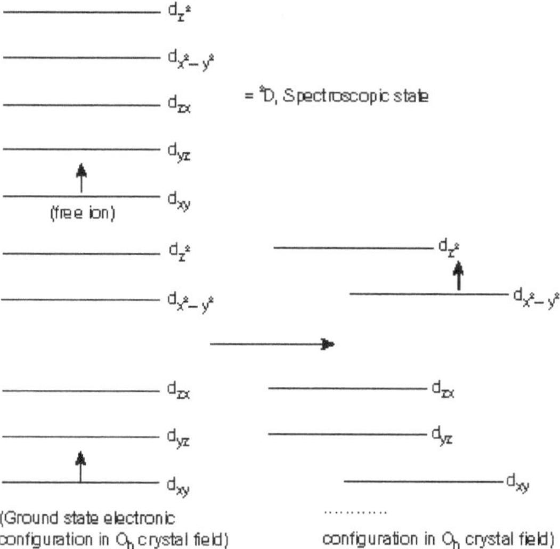

The absorption band corresponds to the followng transition:

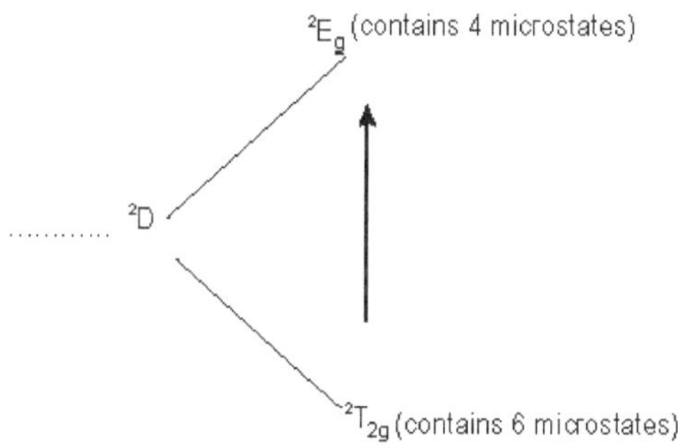

The free ion ground state is the spectroscopic state ^2D, which has 10 microstates. In an octahedral crystal field, the degeneracy of the spectroscopic state is lifted, and the ground term $^2T_{2g}$ will have six microstates (spin degeneracy × orbital degeneracy = 2 × 3 = 6) and the excited state term ^2E_g will have four microstates (spin degeneracy × orbital degeneracy = 2 × 2 = 4). The following may be useful in future discussions:

Group Theoretical Term	Orbital Degeneracy
A term	1
B term	1
E term	2
T term	3

In a tetrahedral environment, the ground state electronic configuration will be $(e)^1$ and the ground term is described by the group theoretical term ^2E. Similarly, the excited state electronic configuration will be $(t_2)^1$ and the excited term is described by the group theoretical term $^2T_2$. One absorption band is predicted corresponding to the electronic transition $(e)^1 \rightarrow (t_2)^1$ and depited as $^2T_2 \leftarrow {}^2E$.

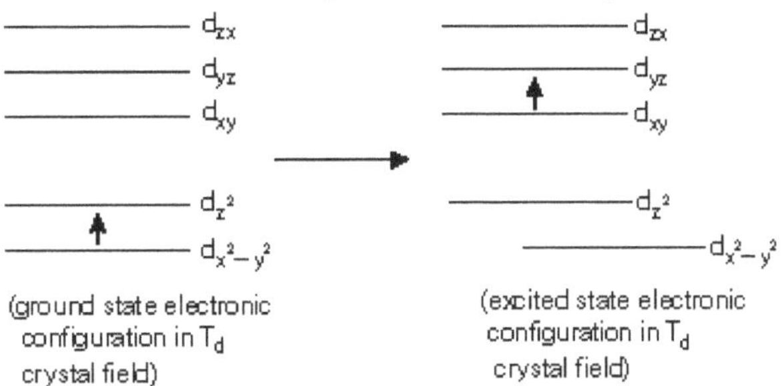

Crystal Field Spectra of $3d^2$ System

In an octahedral environment, the number of microstates for 'free' (absence of any crystal field) d^2 ion is $\dfrac{10 \times 9}{1 \times 2}$, which is equal to 45.

By following the procedure adopted for free C atom, the following spectroscopic states (Table 2.1) can be derived for free d^2 ion.

For example number of microstates in spectroscopic ^1S = spin degeneracy × orbital degeneracy = 1 × 1 = 1; the superscript refers to spin degeneracy or spin multiplicity. S means $L = 0$; therefore, orbital degeneracy $(2L + 1) = 1$. Similarly the number of microstates for other spectroscopic states are listed in the table.

Table 2.1 Number of microstates for different spectoscopic states

Spectroscopic State	(2S+1) Spin Degeneracy	(2L+1) Orbital Degeneracy	Number of Microstates
^1S	1	1	1
^1G	1	9	9
^1D	1	5	5
^3P	3	3	9
^3F	3	7	21

Among the spectroscopic states, that state with maximum spin degeneracy will be the ground state; hence, ^3P or ^3F may be the ground state. However, between ^3P and ^3F, ^3F is the state having maximum orbital degeneracy. Hence, by rule, ^3F is the ground state.

Weak octahedral crystal field   In the weak field limit, crystal field is weak; inter-electronic repulsion is strong.

$$\text{crystal field} < \text{inter-electronic repulsion}$$

Spectroscopic states arise because of inter-electronic repulsion. Superimposing the effect of crystal field upon spectroscopic states, group theoretical terms are obtained. Now, it will be clear how the 45 microstates are distributed among the five spectroscopic states; also, the degeneracy of a spectroscopic state is lifted by a crystal field.

Table 2.2   Effect of inter-electronic repulsion and superimposing effect o crystal field

Spectroscopic State	Group Theoretical Term	Number of Microstates
^1S	^1A_1g	1
^1G	$^1A_1g + {}^1T_1g\ {}^1T_2g + {}^1Eg$	$1+3 + 3 + 2 = 9$
^1D	$^1T_2g + {}^1Eg$	$3 + 2 = 5$
^3P	$3T_{1g}$	9
^3F	$3T_{1g} + 3T_{2g} + 3A_{2g}$	$9 + 9 + 3 = 21$

Strong octahedral crystal field   In the strong field limit, crystal field is strong; inter-electronic repulsion is weak.

$$\text{crystal field} > \text{inter-electronic repulsion.}$$

First, the effect of crystal field is to be considered; then the effect of inter-electronic repulsion is to b superimposed.

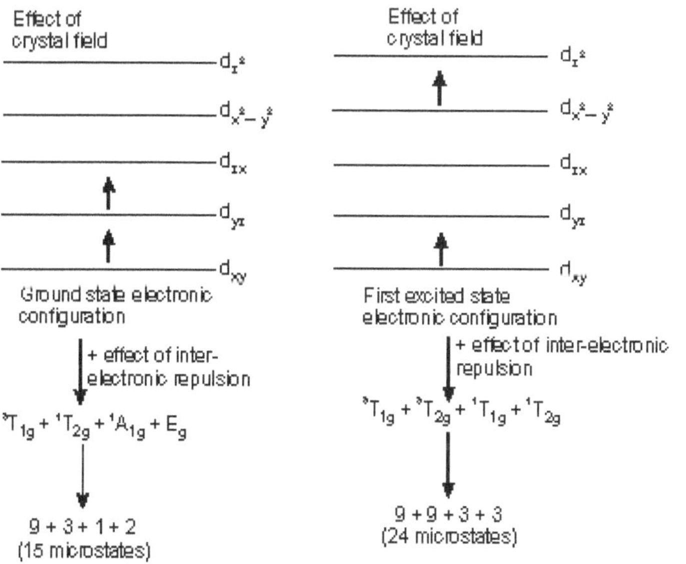

Single electron in t_{2g} set of orbitals can be placed in 6 ways; single electron in e_g set of orbitals can be placed in 4 ways. Hence total number of micro states is equal to $6 \times 4 = 24$.

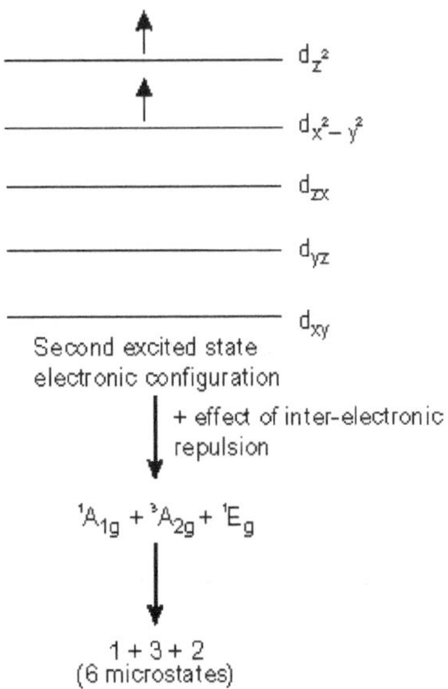

It is seen how group theoretical terms are obtained in the weak crystal field limit and the strong crystal field limit in O_h point group. Whatever be the strength of octahedral crystal field, the same group theoretical terms exist. Since the ground term is a triplet term (spin multiplicity 3), it is sufficient to consider the triplet terms alone to explain the spin allowed in electronic transitions.

where $x = \frac{1}{2}\,(225\,B^2 + 100\,Dq^2 + 180\,Dq\,B)^{1/2}$.

Now energy of v_1 transition $(^3T_{2g} \leftarrow {}^3T_{1g}\,(F))$ is calculated as follows:

$$v_1 = (+2\,Dq) - (7.5\,B - 3\,Dq - x)$$
$$= 5\,Dq - 7.5\,B + x$$

Similarly energy of v_2 transition $(^3A_{2g} \leftarrow {}^3T_{1g}(F))$ is as follows:

$$v_2 = (+12 \text{ Dq}) - (7.5 \text{ B} - 3\text{Dq} - x)$$
$$= 15 \text{ Dq} - 7.5 \text{ B} + x$$

And energy of v_3 transition $(^3T_{1g}(P) \leftarrow {}^3T_{1g}(F))$ is as follows:

$$v_3 = (7.5 \text{ B} - 3\text{Dq} + x) - (7.5 \text{ B} - 3\text{Dq} - x)$$
$$= 2x$$

B is known as Racah parameter. For a free ion, B is always written as B, and for a complex ion, it is denoted as B'.

B'/B is known as nephlauextic ratio (β).

When values of v_1, v_2 and v_3 are given for an octahedral complex of $3d^2$ system, it is easy to calculate the values of 10 Dq and B'. If B value for free ion is given, then the value of nephlauextic ratio β, can be calculated.

Crystal Field Spectra of 3d³ System

Consider an octahedral complex of a $3d^3$ system. Similar to $3d^2$ system we can assign v_1, v_2 and v_3. The ground term will be an A term, as the ground state electronic configuration is $(t_{2g})^3$ and there is only one way of getting maximum spin (getting maximum number of unpaired electrons) for this configuration; orbital degeneracy is one. The following are true for $3d^3$ system in O_h point group.

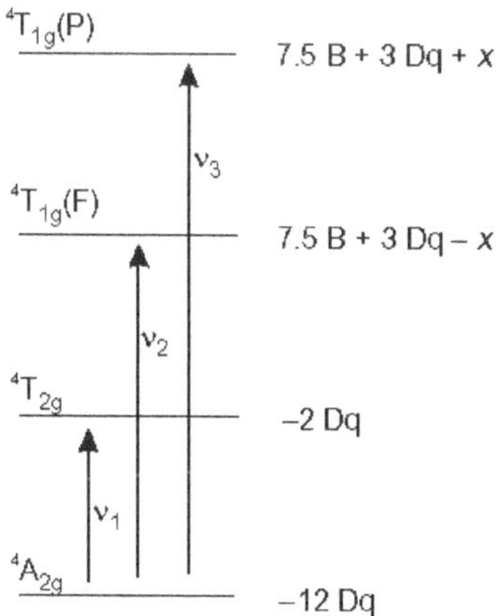

where $x = \frac{1}{2}(225 \text{ B}^2 + 100 \text{ Dq}^2 - 180 \text{ Dq B})^{1/2}$

Now energy of v_1 (transition $^4T_{2g} \leftarrow {}^4A_{2g}$) is calculated as follows:

$$v_1 = (-2 \text{ Dq}) - (-12 \text{ Dq})$$
$$= 10 \text{ Dq}$$

Similarly energy of v_2 transition ${}^4T_{1g}(F) \leftarrow {}^4A_{2g}$ is as follows:

$$v_2 = (7.5\ B + 3\ Dq - x) - (-12\ Dq)$$
$$= 7.5\ B + 15\ Dq - x$$

And energy of v_3 transition ${}^4T_{1g}(P) \leftarrow {}^4A_{2g}$ is:

$$v_3 = (7.5\ B + 3\ Dq + x) - (-12\ Dq)$$
$$= 7.5\ B + 15\ Dq + x$$

When values of v_1, v_2 and v_3 are given for an octahedral complex of $3d^3$ system, it is easy to calculate the values of 10 Dq and B'. If B value for free ion is given, the value of nephlauextic ratio β can be calculated.

Crystal Field Spectra of 3d⁴ System

In the case of $3d^4$ system, the ground term in O_h crystal field depends whether it is a high spin complex or a low spin complex. In the case of a high spin complex the ground term is an E term (orbital degeneracy two) and for the low spin complex the ground term is a T term (orbital degeneracy three).

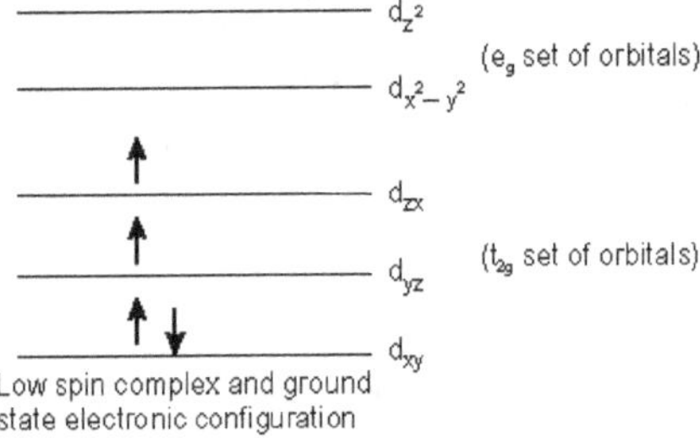

Low spin complex and ground state electronic configuration

Look at how many ways we can get the maximum spin for the $(t_{2g})^4$ configuration. That gives orbital degeneracy and its meaning.

They are given as follows:

Th first one is,

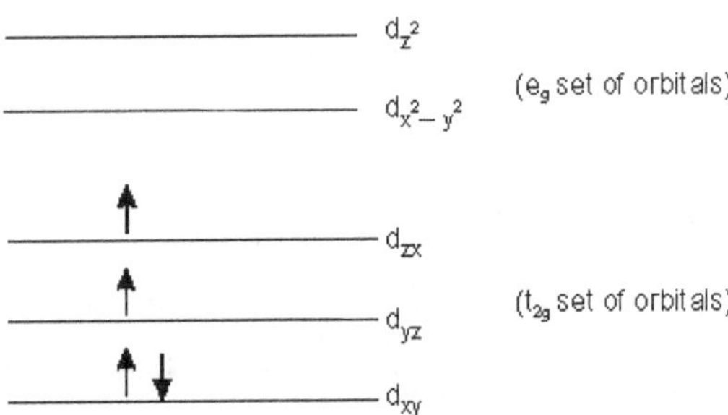

The second one is,

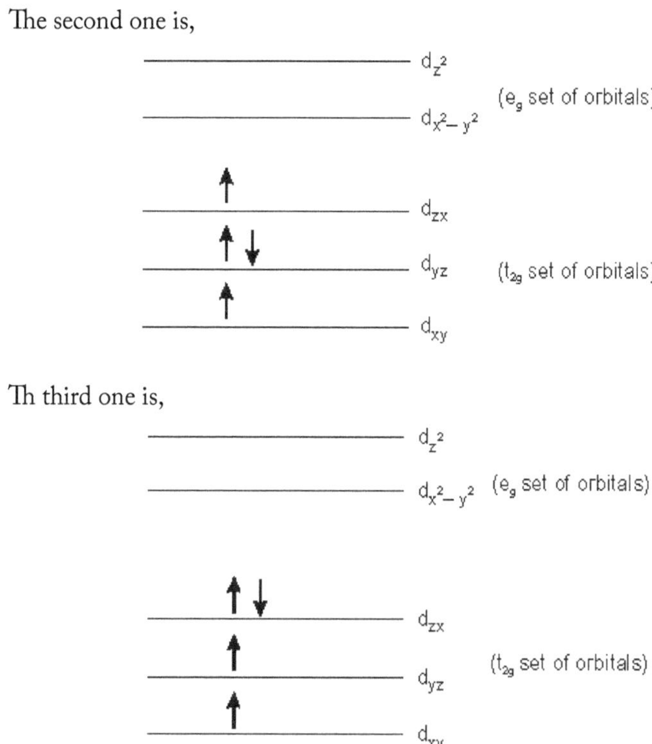

Th third one is,

Also, the spin degeneracy or spin multiplicity of the ground term is 3; there are 2 unpaired electrons, hence S is 1 and the M_S values are 1,0 & -1. Thus the ground term for $3d^4$ low spin system is $^3T_{1g}$. Since many excited triplet terms are available, a richer spectrum is expected. For the high spin complex the ground term is ^5E_g and as there is only one excited quintet term $^5T_{2g}$, one spin-allowed transition, i.e., $^5T_{2g} \leftarrow {}^5E_g$ at an energy of 10 Dq, is expected.

Crystal Field Spectra of $3d^5$ System

In the case of $3d^5$ system, the ground term in O_h crystal field depends on whether it is a high spin complex or a low spin complex. For the high spin complex the ground term is an A term (orbital degeneracy one) and for the low spin complex the ground term is a T term (orbital degenracy three).

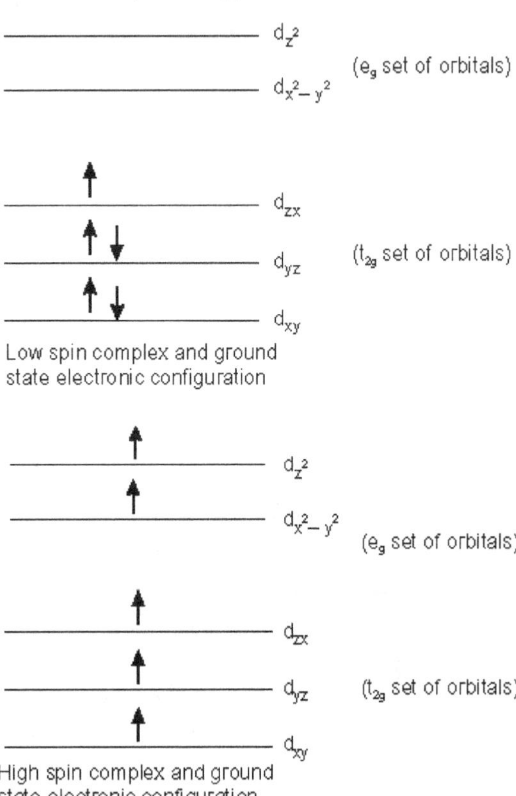

For high spin complex the ground term is $^6A_{1g}$. No sextet-excited term is available. Only spin-forbidden transitions will be responsible for absorption spectrum and hence the absorption will be weak. The ground term for the low spin complex is $^2T_{2g}$ and since excited doublet-terms are available, there will be spin-allowed transitions from this ground state.

Crystal Field Spectra of 3d⁶ System

In the case of $3d^6$ system, the ground term in O_h crystal field depends whether the crystal field is weak or strong. For the high spin complex, the ground term is a T term (orbital degeneracy three) and a for the low spin complex, the ground term is an A term (orbital dgeneracy one).

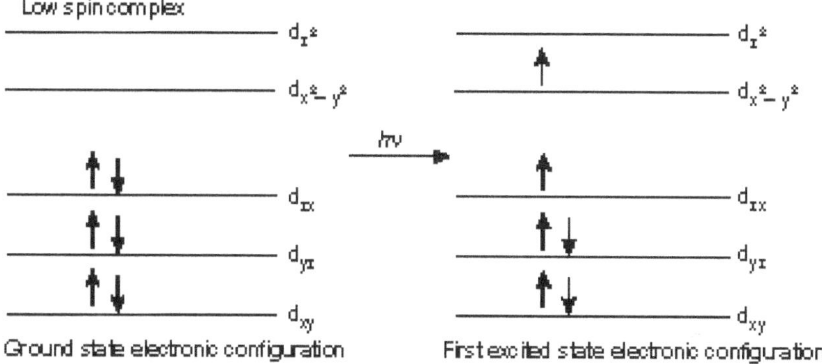

The ground group theoretical term for the low spin octahedral complex is $^1A_{1g}$. The first excited state electronic configuration namely $(t_{2g})^5 (e_g)^1$ spans $^3T_{1g} + {}^3T_{2g} + {}^1T_{1g} + {}^1T_{2g}$. Therefore two spin-allowed transitions are expected and also observed, e.g. low spin Co(II) complexes.

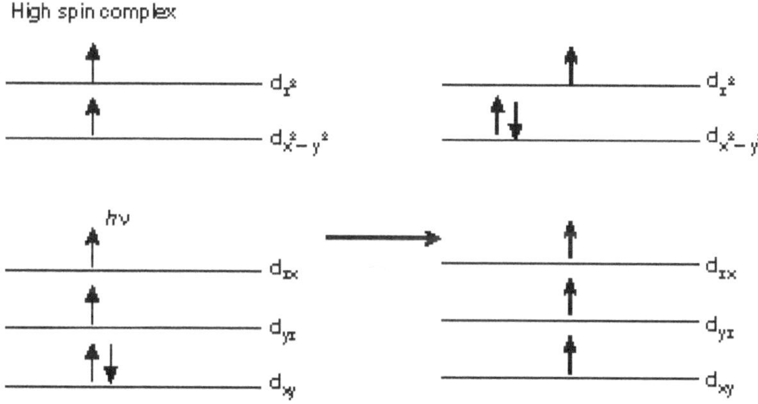

$^5T_{2g}$ term corresponds to the ground state electronic configuration and ^5E_g term corresponds to the excited state electronic configuration; a spin allowed transition $^5E_g \leftarrow {}^5T_{2g}$ is expected to occur at an energy of 10 Dq.

Crystal Field Spectra of 3d⁷ System

Octahedral system   In the case of $3d^7$ system, e.g. Co^{2+}, the ground term in O_h crystal field depends upon the strength of the crystal field, either weak or strong. For the high spin complex, the ground term is a T term (orbital degeneracy three) and for the low spin complex, the ground term is an E term (orbital degeneracy two).

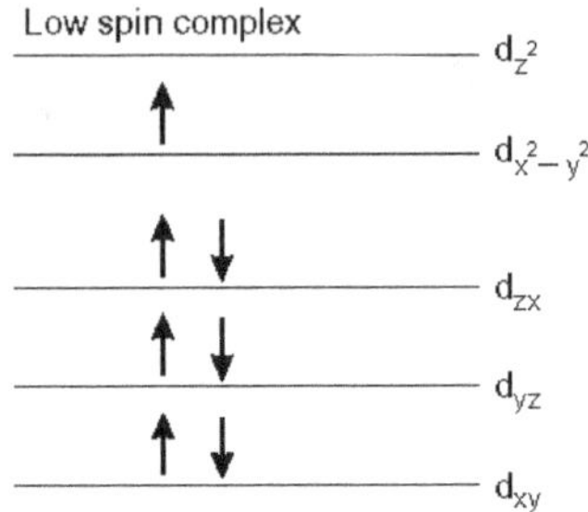

Ground state electronic configuration

The ground group theoretical term for the low spin octahedral complex is ^2E_g. Excited doublet-terms are available; spin-allowed transitions occur.

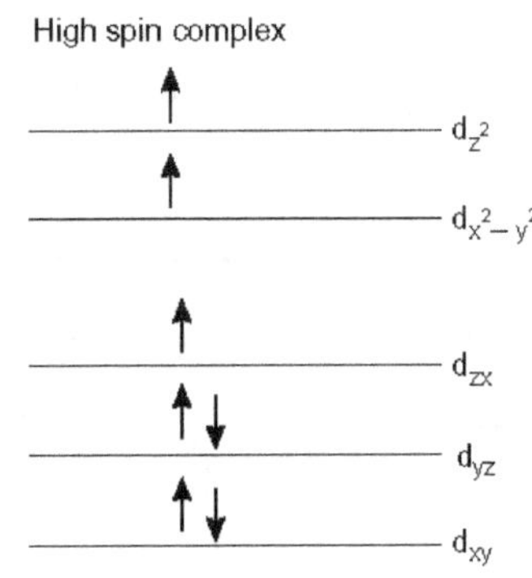

Ground state electronic configuration

The correlation diagram is analogous to that of $3d^2$; three spin-allowed transitions from the spin-free ground state are expected.

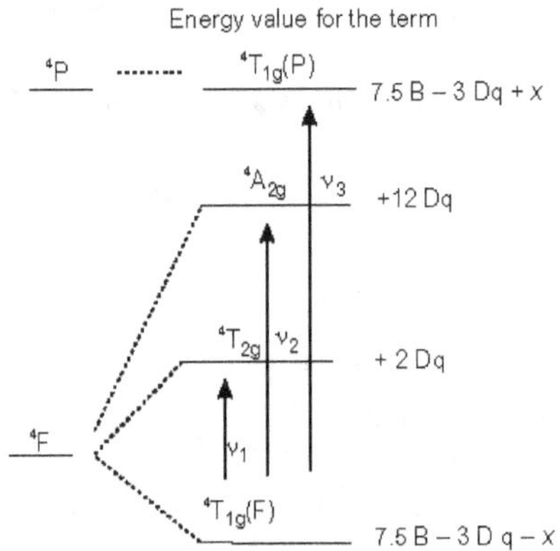

where, $x = \frac{1}{2}(225\,B^2 + 100\,Dq^2 + 180\,Dq\,B)^{1/2}$

Now energy of ν_1 transition $^4T_{2g} \leftarrow {}^4T_{1g}$ (F)) is calculated as follows:

$$\nu_1 = (+2\ Dq) - (7.5\ B - 3\ Dq - x)$$
$$= 5\ Dq - 7.5\ B + x$$

Similarly energy of ν_2 (transition $^4A_{2g} \leftarrow {}^4T_{1g}\ (F)$) is as follows:

$$\nu_2 = (+12\ Dq) - (7.5\ B - 3\ Dq - x)$$
$$= 15\ Dq - 7.5\ B + x$$

And energy of ν_3 (transition $^4T_{1g}(P)) \leftarrow {}^4T_{1g}\ (F)$ is:

$$\nu_3 = (7.5\ B - 3\ Dq + x) - (7.5\ B - 3\ Dq - x)$$
$$= 2x$$

B is known as Racah parameter.

For the free ion, B is always written as B and for a complex it is denoted as B'.

B'/B is known as β, called nephlauextic ratio.

When values of ν_1, ν_2 and ν_3 are given for a high-spin octahedral complex of $3d^7$ system, it is easy to calculate the values of 10 Dq and B'. If B value for free ion is given, the value of nephlauextic ratio β can be calculated.

Tetrahedral system   In the case of tetrahedral Co^{2+}, the ground term is an A term and the correlation diagram is analogous to that of octahedral $3d^3$ and three spin-allowed transition are expected.

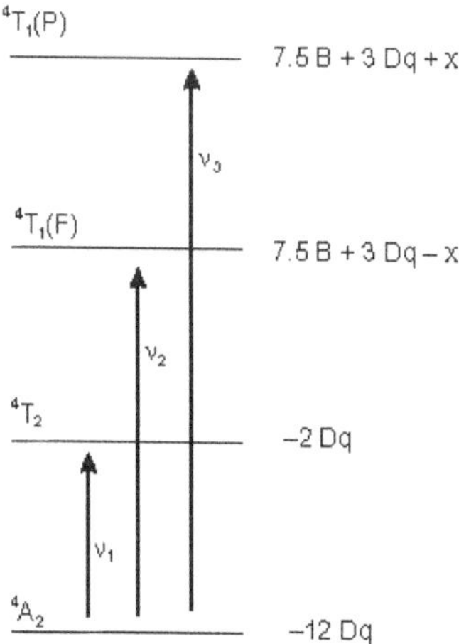

where $x = \frac{1}{2}(225\ B^2 + 100\ Dq^2 - 180\ Dq\ B)^{1/2}$

Now energy of ν_1 transition $^4T_2 \leftarrow {}^4A_2$ is calculated as follows:

$$\nu_1 = (-2\ Dq) - (-12\ Dq)$$
$$= 10\ Dq$$

Similarly energy of v_2 transition $^4T_1(F) \leftarrow {}^4A_2$ is as follows:

$$v_2 = (7.5 \text{ B} + 3 \text{ Dq} - x) - (-12 \text{ Dq})$$
$$= 7.5 \text{ B} + 15 \text{ Dq} - x$$

And energy of v_3 transition $^4T_1(P) \leftarrow {}^4A_2$ is:

$$v_3 = (7.5 \text{ B} + 3 \text{ Dq} + x) - (-12 \text{ Dq})$$
$$= 7.5 \text{ B} + 15 \text{ Dq} + x$$

When values of v_1, v_2 and v_3 are given for a tetrahedral complex of Co^{2+}, it is easy to calculate the values of 10 Dq and B'. If B value for free ion is given, then the value of nephlauextic ratio β can be calculated.

Crystal Field Spectra of 3d⁸ System

Octahedral system   Consider an octahedral complex of a $3d^8$ system, e.g. Ni(II). Similar to $3d^3$ system we can assign v_1, v_2 and v_3. The ground term will be an A term, as the ground state electronic configuration is $(t_{2g})^6 (e_g)^2$. The following are true for $3d^8$ system in O_h point group (no matter, crystal field is weak or strong).

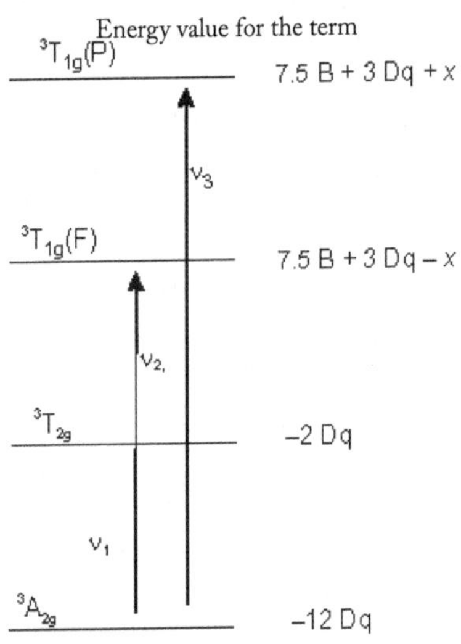

where $x = \frac{1}{2} (225 \text{ B}^2 + 100 \text{ Dq}^2 - 180 \text{ Dq B})^{1/2}$

Now energy of v_1 (transition $^3T_{2g} \leftarrow {}^3A_{2g}$) is calculated as follows:

$$v_1 = (-2 \text{ Dq}) - (-12 \text{ Dq})$$
$$= 10 \text{ Dq}$$

Similarly energy of v_2 (transition $^3T_{1g}(F) \leftarrow {}^3A_{2g}$) is as follows:

$$v_2 = (7.5\ B + 3\ Dq - x) - (-12\ Dq)$$
$$= 7.5\ B + 15\ Dq - x$$

And energy of v_3 (transition $^3T_{1g}(P) \leftarrow {}^3A_{2g}$) is:

$$v_3 = (7.5\ B + 3\ Dq + x) - (-12\ Dq)$$
$$= 7.5\ B + 15\ Dq + x$$

When values of v_1, v_2 and v_3 are given for an octahedral complex of $3d^8$ system (e.g. Ni^{2+}), it is easy to calculate the values of 10 Dq and B'. If B value for free ion is given, then the value of nephlauextic ratio β can be calculated.

Tetrahedral system   In the case of tetrahedral Ni^{2+}, the ground term is a T term and the correlation diagram is analogous to that of octahedral $3d^2$, and three spin-allowed transition are expected.

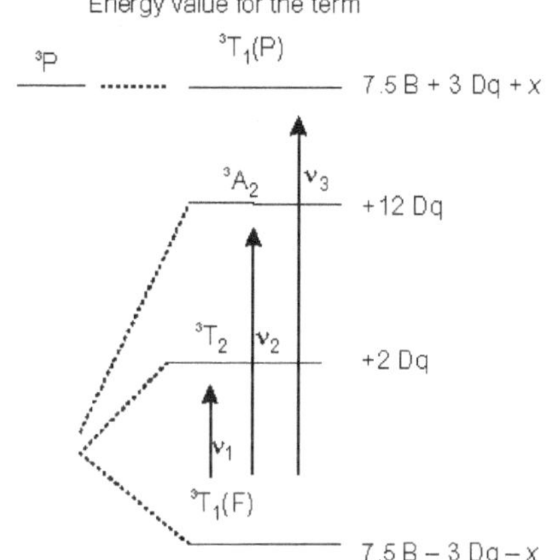

Now energy of v_1 transition $^3T_2 \leftarrow {}^3T_1$ (F) is calculated as follows:

$$v_1 = (+2\ Dq) - (7.5\ B - 3\ Dq - x)$$
$$= 5\ Dq - 7.5\ B + x$$

Similarly energy of v_2 transition $^3A_2 \leftarrow {}^3T_1$ (F) is as follows:

$$v_2 = (+12\ Dq) - (7.5\ B - 3\ Dq - x)$$
$$= 15\ Dq - 7.5\ B + x$$

And energy of v_3 transition $^3T_1(P) \leftarrow {}^3T_1$ (F) is:

$$v_3 = (7.5\ B - 3\ Dq + x) - (7.5\ B - 3\ Dq - x)$$
$$= 2x$$

When values of v_1, v_2 and v_3 are given for a tetrahedral complex of $3d^8$ system, it is easy to calculate the values of 10 Dq and B'. If B value for free ion is given, then the value of nephlauextic ratio β can be calculated.

Crystal Field Spectra of 3d⁹ System

Consider an octahedral complex of a $3d^9$ system (e.g. Cu^{2+}). The electronic configurations of ground state and excited state are as follows in the O_h point group.

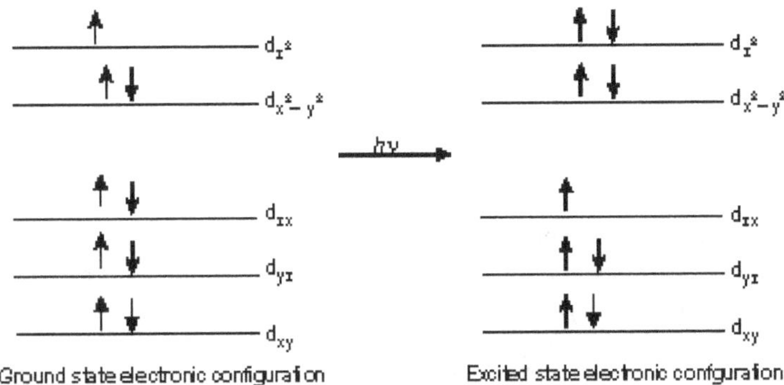

The above transition can be simply represented as: $^2T_{2g} \leftarrow {}^2E_g$

Since in Cu^{2+}, a d^9 system susceptible to Jahn–Teller distortion, the distortion is usually by elongation along z-axis and hence the situation in D_{4th} point group is:

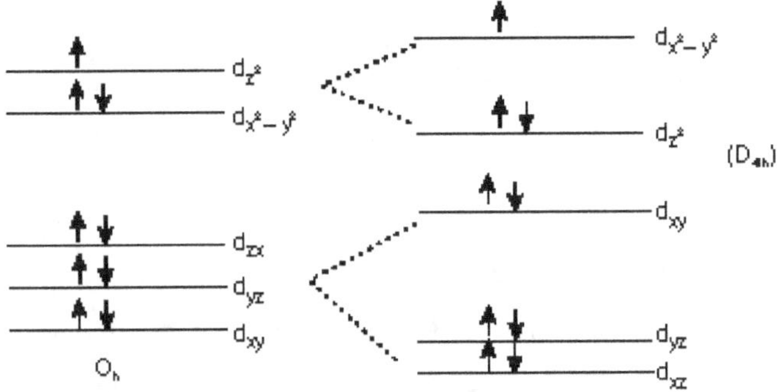

Electronic configurations in D_{4h} point groupis as follows:

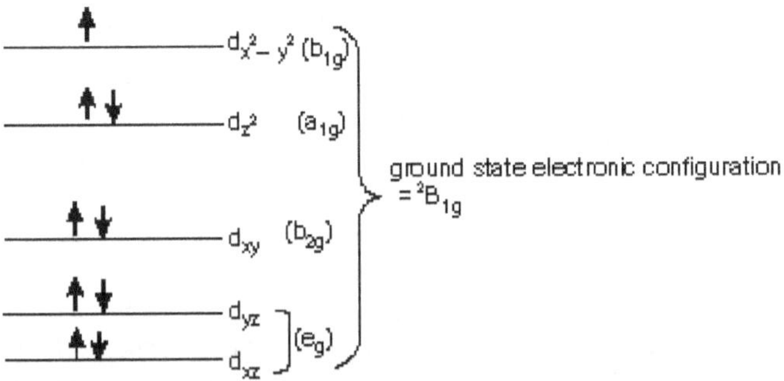

The orbitals $d_{x^2-y^2}, d_{z^2}, d_{xy}$ and (d_{yz} and d_{xz}) can be named with the help of group theory as b_{1g}, a_{1g}, b_{2g} and e_g orbitals respectively. Since the unpaired electrons are present in the b_{1g} orbital in the ground state, the ground term is $^2B_{1g}$.

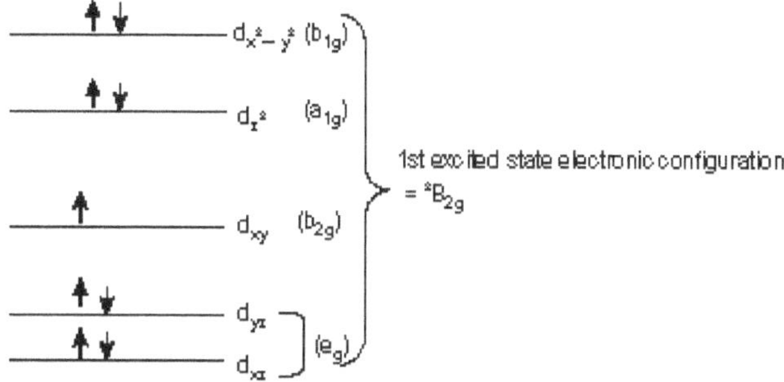

Since the unpaired electrons are present in the b_{2g} orbital in the 1st excited state, the corresponding group theoretical term is $^2B_{2g}$. See what happens in the second excited state. The second excited state can be represented as ^2E_g, and now the various transitions in a D_{4h} point group for Cu^{2+} can be pictorially represented.

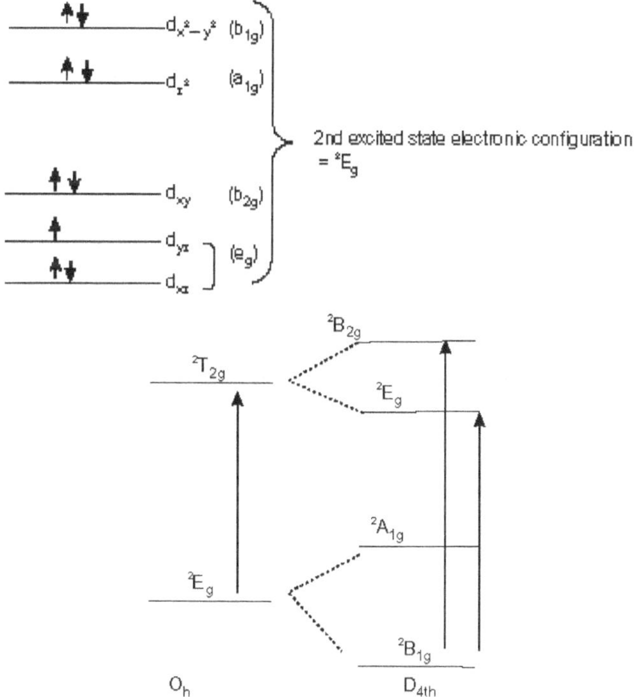

Any Cu^{2+} system in a distorted octahedral environment will exhibit two transitions, viz. $^2B_g \leftarrow {}^2B_{1g}$ and $^2E_g \leftarrow {}^2B_{1g}$ and usually the two transitions will overlap to give a broad band which can be represented as $(^2B_{2g}, {}^2E_g) \leftarrow {}^2B_{1g}$.

Charge Transfer Spectra

Whilst colour in inorganic compounds is generally associated with the presence of a partially filled d shell, there are many such compounds with d^0 or d^{10} configurations which are intensely coloured. Mercury (II) iodide (brick red, d^{10}), the permanganate (VII) ion (intense purple, d^0), bismuth (III) iodide (orange red, $d^{10} s^2$) are few examples. In such cases the colour arises, at least in part, as a consequence of the absorption of light, which occurs when an electron is transferred from an orbital lying **principally** on the ligand to an orbital lying **principally** on the metal or **vice versa**. Such charge transfer (or electron transfer) processes are termed "ligand to metal" or "metal to ligand" charge transfer respectively. Charge transfer processes are commonly of higher quantum energy than crystal field transitions and generally lie in the ultraviolet or far ultraviolet region of the electromagnetic spectrum. However, if the metal is easily oxidizable and the ligand readily reducible, or **vice versa**, then charge transfer transitions may occur in the visible region. The spectra are often very intense and may mask the weaker crystal field transitions.

Charge Transfer Spectra of Halide Complexes

Look at the molecular orbital diagram of an octahedral MX_6 species, where M stands for a metal ion and X stands for a halide ion.

A molecular orbital diagram has been constructed to explain the charge transfer transitions. Combination of the s and p orbitals on each of the six halide ions yields p orbitals transforming as $t_{1g} + t_{2g} + t_{1u} + t_{2u}$ (3 + 3 + 3 + 3 = 12 orbitals), and s orbitals transforming as $a_{1g} + e_g + t_{1u}$ (1+2+3=6 orbitals). The metal orbitals to be considered are n d, (n+1) s and (n+1) p orbitals. In heavy metal hexahalides, the ligand to metal charge transfer transitions v_1, v_2, v_3 and v_4 are observed. Similarly molecular orbital diagrams may be constructed for cyanide complexes and bipyridine related complex systems and the CT can be explained.

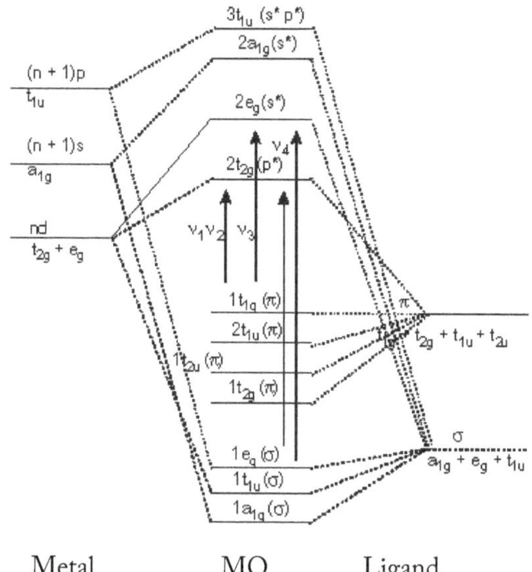

Problems and Solutions

1. Predict the various electronic transitions possible for each of following compounds:

    i. CH_4 ii. CH_3Cl iii. H_2CO iv. Cl

    Solution

    i. $\sigma \rightarrow \sigma^*$ ii. $\sigma \rightarrow \sigma^*$ and $n \rightarrow \sigma$

    iii. $\sigma \rightarrow \sigma^*, \sigma \rightarrow \sigma^*, n \rightarrow \sigma, \sigma \rightarrow \pi$ and $n \rightarrow \sigma^*$ iv. $\sigma \rightarrow \sigma^*$ and $n \rightarrow \sigma$

2. Calculate the absorption maxima of the diene $CH_2 = CH_2$ and explain the deviation from the observed value of 220 nm.

    Calculated value: Base value 253 + substituents (2R) 10 + exo double bond 10 = 273

    The explanation for the discrepancy is the steric strain. The two double bonds are not fully coplanar because the hydrogen atoms of the two methylene groups are very close to each other which results in repulsion. Deviation from coplanarity reduces resonance and lowers $\lambda_{max.}$

3. Acetone absorbs at 279 nm in hexane whereas the value of λ_{max} in water is 264.5 nm. Explain.

    The blue shift results from hydrogen bonding which lowers the energy of n orbitals.

4. Solutions of iodine in hexane is violet while in benzene it is brown. Explain.

   The appearance of brown colour in benzene is due to the formation of charge transfer benzene–iodine complex. The pi electron cloud of benzene interacts with the outer electrons of the iodine molecule.

REVIEW QUESTIONS

1. Which part of a molecule is excited by UV radiation?

2. Which of the following molecules absorb UV radiation? Explain.

   i. heptane

   ii. benzene

   iii. chlorohexane

   iv. ethanol

   v. ammonia

   vi. water

   vii. butadiene

3. Draw a schematic diagram of a double-beam spectrophotometer. Briefly explain the function of each component.

4. What is the principle of a photomultiplier tube?

5. State and explain Beer–Lambert's law.

6. Define the following terms:

   i. bathochromic shift

   ii. hyposchromic shift

   iii. hyperchromic shift

   iv. hypochromic shift

   v. chromophore

   vi. auxochrome

7. State the selection rules for electronic spectroscopy.

8. Explain why an aqueous solution of manganese sulphate is almost colourless.

9. A solution shows a transmittance of 20% when taken in a cell of 2.5 cm thickness. Calculate its concentration if the molar absorption coefficient is 12000 $dm^2 \, mol^{-1} \, cm^{-1}$.

10. Briefly describe any two applications of spectrophotometric technique.

11. The maximum absorption wavelength for 1,3 butadiene is higher than that for ethylene. Why?

12. What is the difference between a single-beam and a double-beam spectrophotometer?

13. What value of absorbance corresponds to 45.0% T?

14. State the difference between transmittance, absorbance and molar absorptivity. Which one is proportional to concentration?

15. Arrange the following in the increasing order of their λ_{max}.

    i.   anthracene

    ii. butadiene

    iii. ethylene

    iv. napthalene

16. Explain why aminoazobenzene is yellow while its acid solution is violet in colour.

17. Predict the various transitions possible in the following compounds:

    i. CH_4

    ii. CH_2Cl

    iii. $H_2C\ O$

    iv. Cl_2.

18. What are the various detectors used in UV–visible spectrophotometers?

19. The wavelength of maximum absorption for methyl chloride is 173 nm while that for methyl iodide is 253 nm. Explain.

20. Predict the transitions involved in saturated aldehydes and ketones.

21. The octahedral hydrated cobalt (II) complex, $[Co(H_2O)^{2+}$ appears pink in acid solution, while tetrahedral cobalt (III) complex of the type $[CoX_4]^{2-}$ appears blue. Account for.

22. The complex $[Ti(H_2O)_6]^+$ absorbs green and yellow light. Explain.

23. Why do absorptions appear as bands instead of sharp lines in UV spectrum?

24. Calculate the transmittance and absorbance of a solution containing 5.2 ppm of the solute.

25. Why do free radicals and odd electron molecules absorb in the visible region?

Chapter 3

INFRARED SPECTROSCOPY

Introduction

The term 'infrared (IR)' covers a range of electromagnetic spectrum between 0.78 and 1000 μm. In the context of infrared spectroscopy, wavelength is measured in 'wavenumbers', whose unit is cm^{-1}.

$$\text{Wavenumber} = \frac{1}{\text{Wavelength (in centimetres)}}$$

It is useful to divide the infrared region into three sections: near, mid and far infrared.

Region	Wavelength range (μm)	Wavenumber range (cm^{-1})
Near	0.78–2.5	12800–4000
Middle	2.5–50	4000–200
Far	50–1000	200–10

The most useful IR region lies between 4000 and 670 cm^{-1}.

Instrumentation

An IR spectrometer consists of three basic components: radiation source, monochromator and detector. A common radiation source for IR spectrometer is an inert solid heated electrically to 1000–1800°C. Three popular types of sources are Nernst glower (constructed of rare earth oxides), Globar (constructed of silicon carbide) and nichrome coil. They produce continuous radiations but with different radiation energy profiles.

The monochromator is a device used to disperse a broad spectrum of radiation and provide a continuous calibrated series of electromagnetic energy bands of determinable wavelength or frequency range. Prisms or gratings are the dispersive components used in conjunction with variable-slit mechanisms, mirrors and filters. For example, a grating rotates to focus a narrow band of frequencies on a mechanical slit. Narrower slits enable the instrument to better distinguish more closely spaced frequencies of radiation, resulting in better resolution. Wider slits allow more light to reach the detector and provide better system sensitivity. Thus, certain compromise is exercised in setting the desired slit width.

Most detectors used in dispersive IR spectrometers can be categorized into two types: thermal and photon detectors. Thermal detectors include thermocouples, thermistors and pneumatic devices (Golay detectors). They measure the heating effect produced by infrared radiation. A variety of physical property changes can be quantitatively determined: expansion of a non-absorbing gas (Golay detector), electrical resistance (thermistor), and voltage at junction of dissimilar metals (thermocouple).

Photon detectors rely on the interaction of IR radiation and a semiconductor material. Non-conducting electrons are excited to a conducting state. Thus, a small current or voltage can be generated. Thermal detectors provide a linear response over a wide range of frequencies but exhibit slower response times and lower sensitivities than photon detectors. A typical dispersive spectrometer is shown in Figure 3.1.

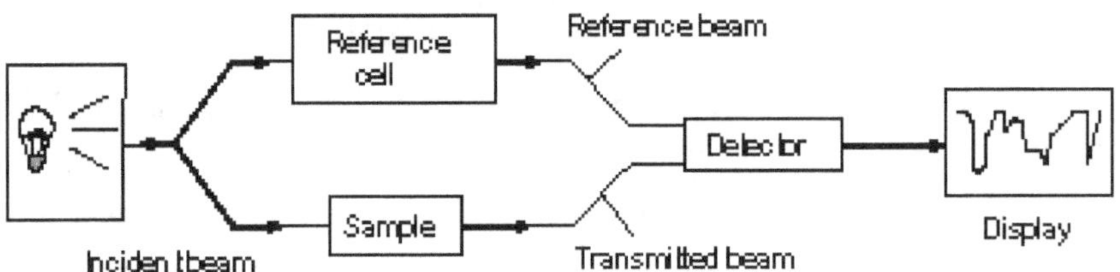

Figure 3.1   Dispersive spectrometer

If the sample is a liquid, it can be tested straightaway. If it is a solid, then it is ground to a fine powder and mixed with a few drops of liquid paraffin (Nujol) to form a paste. A thin layer of the liquid or paste is then spread between two sodium chloride plates and placed in the machine. Sodium chloride is used because it does not absorb strongly in the infrared region whereas glass does; however, it does dissolve readily in water, and so it must be cleaned with CH_2Cl_2. The sample is then placed in the machine.

The reference cell is an identical rock salt plate with a similar amount of Nujol (if used in sample). The detector then compares the two beams it receives, and can remove any peaks due to Nujol or the plates. A detector measures the amount of radiation at various wavelengths that is transmitted by the sample. This information is recorded on a chart, where the per cent of incident light that is transmitted through the sample (% transmission) is plotted against wavelength in microns (μm) or frequency (cm^{-1}) (Figure 3.2).

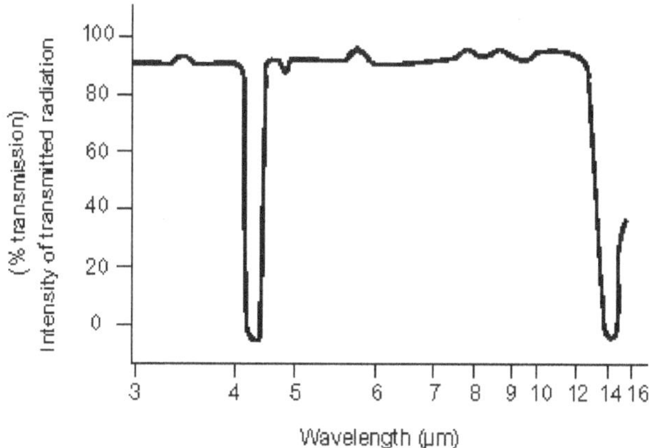

Figure 3.2   Plot of transmitted radiation versus wavelength

Experimental Infrared Spectra

Cells and Gas Phase Spectra

These types of spectra were more a curiosity and of theoretical interest until the introduction of the combined techniques of gas chromatography–Fourier transfer infrared spectroscopy (GC–FTIR). The main advantages of this method are that the spectra can be obtained on micrograms of material, and the spectra do not show the effects of interactions between molecules that are characteristic of condensed phase spectra. These spectra are usually obtained at elevated temperatures. Condensed phase spectra, however, will continue to be important because of the fact that many compounds do not survive injection into a gas chromatograph. Currently, most frequency correlations for various functional groups are reported for the condensed phase spectra. Frequencies observed in the gas phase are usually slightly higher than those observed for the same functional group in the condensed phase.

Gas phase spectra can also be taken at room temperature. All that is needed is a sample with a vapour pressure of several millimeters and a path length of about a decimetre (10 cm). Cells with NaCl or KBr windows are commercially available or can be built easily. Crystals of KBr are transparent from 4000–250 cm^{-1} and are perfectly acceptable for most applications. They have the disadvantage of being hygroscopic and must be stored in a desiccator. Cells of sodium chloride are transparent from 4000–600 cm^{-1}, less expensive and less hygroscopic. These cells are also acceptable for routine spectra.

Cells and Condensed Phase Spectra

Condensed phase spectra may be taken as a solid or as a liquid. The comparison of the same sample in the liquid and solid phase will differ. However, the major differences observed will be in the fingerprint region. In cases where infrared spectroscopy is used as a criterion of identity, the spectra under comparison should be obtained under identical experimental conditions. Liquid phase spectra are the easiest to obtain. All that is needed are two polished discs of NaCl or KBr. A thin film is prepared by depositing a drop of the liquid between the two plates and mounting them in the beam of the spectrometer. This is referred to as a neat liquid. Glass is not used in infrared spectroscopy because of the strong absorptions due to the Si—O group.

Spectra of solids can be obtained in a variety of ways. The method of choice varies depending on the physical properties of the material under consideration. Several methods that can be used satisfactorily along with the limitations and advantages of each are listed below.

Thin film In order to obtain an infrared spectrum of a solid, it is necessary to get light, mainly infrared, through the sample. There may be several ways to achieve this. We shall outline one method that has proven successful in the past. A thin layer of a solid is deposited as a solution on an infrared cell and the solvent is allowed to evaporate. Solvents such as $CHCl_3$, CH_2Cl_2 and CCl_4 have been frequently used. The solid sample should have an appreciable solubility in one of these solvents. A drop of solution left to evaporate will deposit a thin film of crystal that will often transmit sufficient light to provide an acceptable infrared spectrum. However, this method suffers from the disadvantage that a spectrum of the solvent must also be run to determine whether all of the solvent has evaporated.

Nujol mull A mull is a suspension of a solid in a liquid. In this arrangement, light can be transmitted through the sample to afford an acceptable infrared spectrum. The commercial sample of nujol, or mineral oil, which has a long chain hydrocarbon, is often used for this purpose. Most solids do not dissolve in this medium but can be ground up in its presence. A small mortar and pestle is used for this purpose. If the grinding process gives rise to small particles of solid with diameters roughly the same as the wavelength of the infrared radiation being used, (2–5 microns), these particles will scatter light rather than transmit it. If you find this type of band distortion with either a nujol mull or a KBr pellet, simply continue grinding the sample until the particles become finer.

The major disadvantage of using a nujol mull is that the information in the C—H stretching region is lost because of the absorptions of the mulling agent. To eliminate this problem, it is necessary to run a second spectrum in a different mulling agent that does not contain any C—H bonds. Typical mulling agents that are used for this purpose are perfluoro- or perchlorohydrocarbons. Examples include perchlorobutadiene, perfluorokerosene and perfluorohydrocarbon oil.

KBr pellets A KBr pellet is a dilute suspension of a solid in a solid. It is usually obtained by first grinding the sample in anhydrous KBr at a ratio of approximately one part sample to 100 parts KBr. Although it is best to weigh the sample (1 mg) in the KBr (100 mg), it is possible with some experience to use your judgment in assigning proportions of sample to KBr. The mixture is then ground in an apparatus called Wiggle-Bug, frequently used by dentists to prepare amalgams. The ground sample mixture is placed on a steel plate containing a paper card with a hole punched in it. The paper should have the thickness and consistency of a postcard and the hole is positioned such that it will lie in the infrared beam when placed on the spectrometer. The sample is placed in the hole making sure that it also overlaps the paper card. A second steel plate is placed over the sample, and the steel sandwich is now placed in a hydraulic press and subjected to pressures of 15000 psi for 20 seconds. The paper card is removed following decompression, and what you see is a KBr pellet that is reasonably transparent to both visible light and infrared radiation. Some trial and error may be necessary before quality pellets can be obtained routinely. Samples that are not highly crystalline do not produce quality pellets. However, good-quality spectra can be obtained on most samples. The only limitation in KBr is that it is hygroscopic. Hence, it may be a good idea to obtain a spectrum run as a nujol mull on the sample as well. The two spectra should be very similar, and since nujol is a hydrocarbon and has no affinity for water, any absorption in nujol between 3400 and 3600 cm⁻¹ can be attributed to the sample and not to the absorption of water by KBr.

Vibrational Frequencies and Normal Vibrational Modes

In a molecule, the atoms are not held rigidly apart. Instead, they can move as if they are attached by a spring of equilibrium separation, R_e. This bond can easily bend or stretch. If the bond is subjected to infrared radiation of a specific frequency (between 300 and 4000 cm⁻¹), it will absorb energy, and the bond will move from the lowest vibrational state to the next highest state. In a simple diatomic molecule, there is only one direction of vibrating and stretching. This means that there is only one band of infrared absorption.

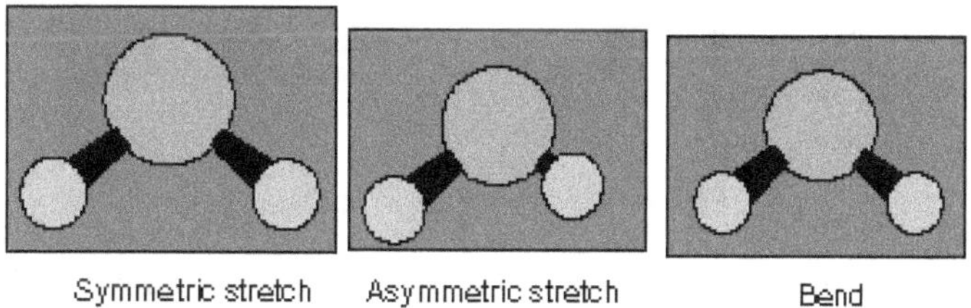

Figure 3.3 Modes of vibrations of water molecule

Weaker bonds require less energy, as if the bonds are springs of different strengths. If there are more atoms, there will be more bonds, and therefore more modes of vibrations. This will produce a more complicated spectrum. For a linear molecule with n atoms, there are $3n - 5$ vibrational modes; if it is non-linear, it will have $3n - 6$ modes. For example, water (H_2O) has three molecules and is non-linear; therefore, it has $(3 \times 3) - 6 = 3$ modes of vibration (Figure 3.3).

Carbon dioxide, a linear molecule, has $(3 \times 3) - 5 = 4$ vibrations. These vibrational modes, shown in Figure 3.4, are responsible for the "greenhouse effect" in which heat radiated from earth is absorbed (trapped) by CO_2 molecules in the atmosphere. The arrows indicate the directions of motion. Vibrations labelled A and B represent the stretching of the chemical bonds, one in a symmetric (A) fashion in which both $C = O$ bonds lengthen and contract together (in phase), and the other in an asymmetric (B) fashion in which one bond shortens while the other lengthens. The asymmetric stretch (B) is infrared **active** because there is a change in the molecular dipole moment during its vibration. To be **active** means that absorption of a photon to excite the vibration is **allowed** by the rules of quantum mechanics.

The infrared "selection rule" states that, for a particular vibrational mode to be observed (active) in the infrared spectrum, the mode must involve a change in the dipole moment of the molecule.

Infrared radiation at 2349 cm^{-1} (4.26 μm) excites this particular vibration. The symmetric stretch is not infrared-active, and so this vibration is not observed in the infrared spectrum of CO_2. The two equal-energy bending vibrations in CO_2 (C and D in Figure 3.4) are identical except that one bending mode is in the plane of the paper, and the other is out of the plane. Infrared radiation at 667 cm^{-1} (15.00 μm) excites these vibrations. Another way of illustrating the out-of-plane mode is to place circled + or − signs on the atoms, signifying motion above or below the plane of the paper, respectively. It takes more energy (shorter wavelengths, higher frequencies) to excite the stretching vibration than the bending vibration.

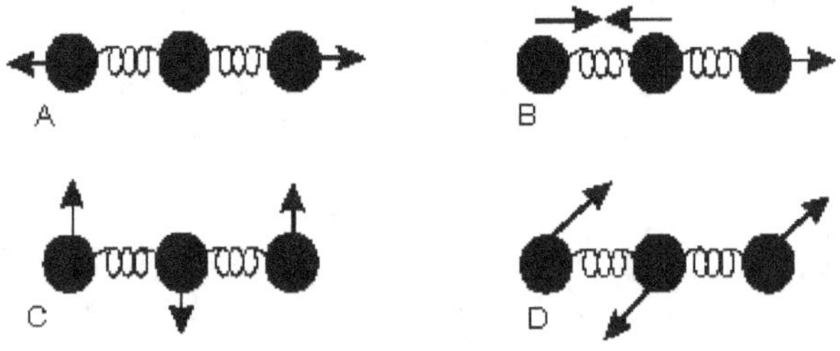

Figure 3.4   Vibrations of CO_2

In addition to bond stretching and bond bending, more complicated molecules vibrate in rocking and twisting modes, which arise from combinations of bond bending in adjacent portions of a molecule. Torsions involve changes in dihedral angles. This type of mode is analogous to twisting the lid off the top of a jar. No bonds are stretched, and no bond angles change, but the spatial relationship between the atoms attached to each of two adjacent atoms will change. The torsional mode for ethane is illustrated in Figure 3.5.

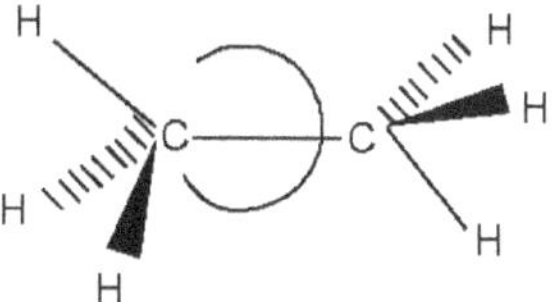

Figure 3.5   Torsional mode of ethane

Without going into more details at this point, let us note some general trends. The stronger the bond, the more the energy required to excite the stretching vibration. This is seen in organic compounds where stretches for triple bonds such as $C \equiv C$ and $C \equiv N$ occur at higher frequencies than stretches for double bonds ($C = C$, $C = N$, $C = O$), which are in turn at higher frequencies than single bonds ($C—C$, $C—N$, $C—H$, $O—H$ or $N—H$). The heavier the atom, the lower the frequencies of vibrations that involve that atom. The characteristic regions for common infrared stretching and bending vibrations are given in Figure 3.6 and Table 3.1.

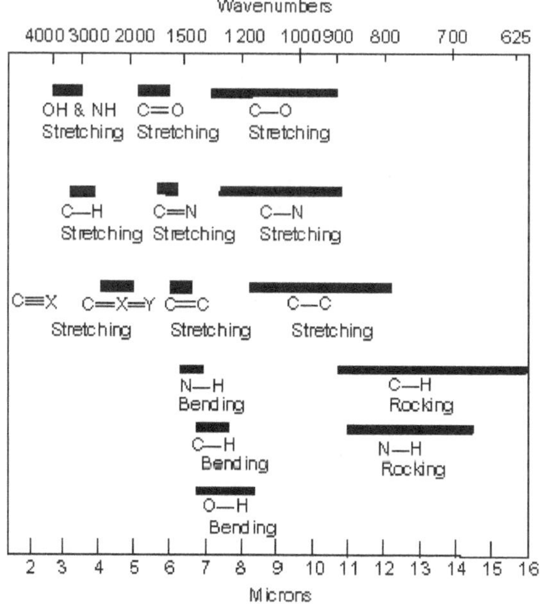

Figure 3.6   Common infrared stretching and bending vibrational frequencies

An important restriction is that the molecule will absorb radiation only if the vibration is accompanied by a change in the dipole moment of the molecule. A dipole occurs when there is a difference of charge across a bond. If the two oppositely charged molecules get closer or further apart as the bond bends or stretches, the moment will change. To calculate the frequency of light absorbed, we use the Hooke's law:

$$\nu_{osc} = \frac{1}{2\pi} \sqrt{k \frac{m_1 + m_2}{m_1 m_2}}$$

where, k is the force constant indicating the strength of the bond; m_1 and m_2 are the masses of the two atoms.

By looking at this equation, we understand that, if there is a high value of k, i.e., the bond is strong, it absorbs a higher frequency of light. So, a C $=$ C) double bond would absorb a higher frequency of light than a (C—C) single bond. Also, the larger the two masses, the lower the frequency of light absorbed.

The rocking, scissoring, wagging and twisting modes are depicted in Figure 3.7

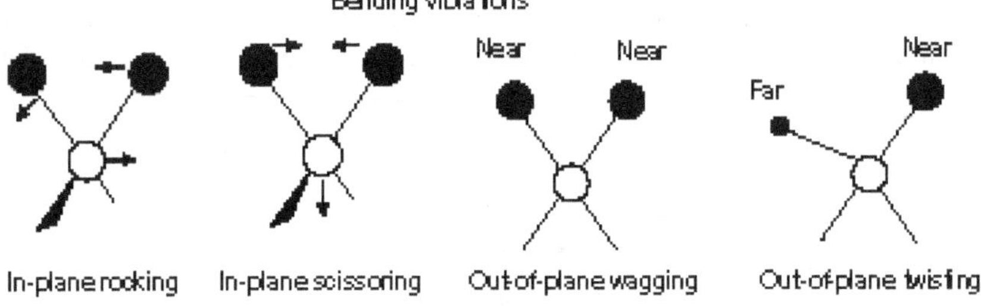

Figure 3.7   Bending vibrations

Vibrational Coupling

Vibrational coupling is the interaction between vibrating bonds that are joined to a single, central atom. It is influenced by a number of factors.

* Strong coupling of stretching vibrations occurs when there is a common atom between the two vibrating bonds.

* Coupling of bending vibrations occurs when there is a common bond between vibrating groups.

* Coupling between a stretching vibration and a bending vibration occurs when the stretching bond is one side of an angle varied by bending vibration.

* Coupling is greatest when the coupled groups have approximately equal energies.

* No coupling is seen between groups separated by two or more bonds.

Table 3.1 A summary of principle infrared bands and their assignments

Functional group	Type		Frequencies (cm^{-1})	Peak intensity
C—H	sp^3 hybridized	R$_3$C—H	2850–3000	M(sh)
	sp^2 hybridized	=CR—H	3000–3250	M(sh)
	sp hybridized	C—H	3300	M-S(sh)
	Aldehyde C—H	H—(C=O)R	2750, 2850	M(sh)
N—H	Primary amine, amide	RN—H$_2$, RCON—H$_2$	3300, 3340	S, S(br)
	Secondary amine, amide	RNR—H, RCON—HR	3300–3500	S(br)
	Tertiary amine, amide	RN(R$_3$), RCONR$_2$	None	
O—H	Alcohols, phenols	Free O—H	3620–3580	W(sh)
		hydrogen bonded	3600–3650	S(br)
	Carboxylic acids	R(C=O)O—H	3500–2400	S(br)
CN	Nitriles	RCN	2280–2200	S(sh)
CC	Acetylenes	R—CC—R	2260–2180	W(sh)
		R—CC—H	2160–2100	M(sh)
C=O	Aldehydes	R(C=O)H	1740–1720	S(sh)
	Ketones	R(C=O)R	1730–1710	S(sh)
	Esters	R(COO)R	1750–1735	S(sh)
	Anhydrides	R(COOCO)R	1820, 1750	S, S(sh)
	Carboxylates	R(COO)H	1600, 1400	S, S(sh)
C=C	Olefins	R$_2$C=CR$_2$	1680–1640	W(sh)

R is an aliphatic group.

Factors Influencing the Location and Number of Peaks

Before we begin a detailed analysis of the various peaks observed in the functional group region, it might be useful to mention some of the factors that can influence the location and number of peaks obtained in infrared spectroscopy. Theoretically, the number of fundamental vibrations or normal modes available to a polyatomic molecule made up of n atoms is given by $3n - 5$ for a totally linear molecule and $3n - 6$ for all others. By normal mode or fundamental vibration, we mean the simple independent bending or stretching motions of two or more atoms which, when combined with all the normal modes associated with the remainder of the molecule, will reproduce the complex vibrational dynamics associated with the real molecules. Normal modes are determined by a normal coordinate analysis. If each of these fundamental vibrations were to be observed, we would expect either $3n - 5$ or $3n - 6$ infrared bands. Some factors decrease the number of bands observed and others cause an increase in this number.

Factors that increase the number of infrared bands   We have earlier mentioned *overtones*, which are absorptions of energy caused by a change of 2 rather than 1 in the vibrational quantum number. While overtones are usually forbidden transitions and, therefore, are weakly absorbing, they do give rise to more bands than expected. Overtones are easily identified by the presence of a strongly absorbing fundamental transition at slightly more than half the frequency of the overtone. Occasionally, combination bands are also observed in the infrared. These bands, as their name implies, are absorption bands observed at frequencies such as $1 + 2$ or $1 - 2$, where 1 and 2 refer to fundamental frequencies. Other combinations of frequencies are also possible. The symmetry properties of the fundamentals play a role in determining which combinations are observed. Fortunately, combination bands are seldom observed in the functional group region of most polyatomic molecules, and the presence of these bands seldom causes a problem in identification. The splitting of bonds in infrared can be caused by Fermi resonance. While a discussion of Fermi resonance is beyond the scope of this chapter, this splitting can be observed whenever two fundamental motions or a fundamental and a combination band have nearly the same energy (i.e., 1 and 2 or 1 and 2 + 3). In this case, the two levels split each other. One level increases in energy while the other decreases. In order to observe Fermi resonance, other symmetry properties of these vibrations must be satisfied in addition to the requirement that a near coincidence of energy levels occurs. As a consequence, Fermi resonance bands are not frequently encountered.

Factors that decrease the number of infrared bands   There are several factors that decrease the number of infrared bands observed. Symmetry is one of the factors that can significantly reduce the number of bands observed in the infrared. If stretching a bond does not cause a change in the dipole moment, the vibration will not be able to interact with the infrared radiation, and the vibration will thus be infrared inactive. Other factors include the near coincidence of peaks that are not resolved by the spectrometer and the fact that only a portion of the infrared spectrum is usually accessed by most commercial infrared spectrometers.

Spectral features of major functional groups and interpretation of infrared spectra

Carbon–Hydrogen Stretching Frequencies

Consider the carbon–hydrogen stretching frequencies. If we assume that all C—H stretching force constants are similar in magnitude, we would expect the stretching frequencies of all C—H bonds to be similar. This expectation is based on the fact that the mass of a carbon atom and whatever else is attached to the carbon is much larger than the mass of hydrogen. The reduced mass for vibration of a hydrogen atom would be approximately the mass of hydrogen atom that is independent of the structure. All C—H stretching frequencies are observed at approximately 3000 cm^{-1}, exactly as expected. Fortunately, the force constants do vary with structure in a fairly predictable manner, and therefore it is possible to differentiate between different types of C—H bonds. You may recall from a foundation course on organic chemistry that the C—H bond strength would increase as the *s* character of the C—H bond increases. Some typical values are given in Table 3.2 for various hybridization states of carbon. Bond strength and bond stiffness deal with different properties. Bond strength measures the depth of the potential energy well associated with a C—H bond. Bond stiffness is a measure of how much energy it would take to compress or stretch a bond. While these are different properties, the stiffer bond is usually associated with a deeper potential energy surface. Table 3.2 shows that increase in the bond strength would increase the C—H bond stretching frequency.

Table 3.2   Carbon-hydrogen bond strengths as a function of hybridization

Type of C—H bond	Bond strength	IR frequency	kcal/mol/cm
sp^3 hybridized C—H	$CH_3CH_2CH_2$—H	99	<3000
sp^2 hybridized C—H	CH_2=CH—H	108	>3000
sp hybridized C—H	HCC—H	128	3300

C–H sp³ hybridization   Methyl groups, methylene groups and methine hydrogens on sp^3 carbon atoms absorb between 2850 and 3000 cm^{-1}. While it is sometimes possible to differentiate between these types of hydrogen, the beginners should probably avoid this type of interpretation. It should be pointed out, however, that molecules that have local symmetry will usually show symmetric and asymmetric stretching frequencies. Take, for example, a CH_2 group. It is not possible to isolate an individual frequency for each hydrogen. The two hydrogens would couple and show two stretching frequencies, a symmetric stretching frequency in which stretching and compression of both hydrogens occur simultaneously, and an asymmetric stretching frequency in which the stretching of one hydrogen is accompanied by the compression of the other. While these two motions occur at different frequencies, both will be found between 2850–3000 cm^{-1} envelope. Similarly, for a CH$_3$ group, symmetric and asymmetric vibrations are observed. This behaviour is observed whenever this type of local symmetry is present. These peaks are usually sharp and are of medium intensity. Considerable overlapping of several of these bands would result in absorption that is fairly intense and broad in this region.

C–H sp^2 hybridization   Hydrogens attached to sp^2 carbons absorb at 3000–3250 cm^{-1}. Both aromatic and vinylic carbon–hydrogen bonds are found in this region. These peaks are usually sharp and of low to medium intensity.

C–H sp hybridization   Hydrogens attached to sp carbons absorb at 3300 cm^{-1}. These peaks are usually sharp and of medium to strong intensity.

C–H aldehydes   The C—H bond of an aldehyde is one additional type of C—H stretch. The C—H stretching frequency appears as a doublet, at 2750 and 2850 cm^{-1}. You may ask why the stretching of a single C—H bond in an aldehyde leads to two bands. The splitting of C—H stretching frequency into a doublet in aldehydes is due to the phenomenon called Fermi resonance. It is believed that the aldehyde C—H stretch is in Fermi resonance with the first overtone of the C—H bending motion of the aldehyde. The normal frequency of the C—H bending motion of an aldehyde is at 1390 cm^{-1}. As a result of this interaction, one energy level drops to *ca.* 2750 and the other increases to *ca.* 2850 cm^{-1}. Only one C—H stretch is observed for aldehydes that has the C—H bending motion significantly shifted from 1390 cm^{-1}.

C–H exceptions   In summary, it is possible to identify the type of hydrogen based on hybridization by examining the infrared spectra in the 3300 to 2750 cm^{-1} region. However, there are some exceptions to the rules we just outlined. Cyclopropyl hydrogens, which are formally classified as sp^3 hybridized, actually have *s* character more than 25%. Carbon–hydrogen frequencies greater than 3000 cm^{-1} are observed for these stretching vibrations. Halogen substitution can also affect the C—H stretching frequency. The C—H stretching frequencies of hydrogens attached to a carbon bearing halogen substitution can also be shifted to above 3000 cm^{-1}. The last exception we will mention is an interesting case in which the force constant is increased because of steric interactions. In the case of infrared spectrum of tri-t-butylcarbinol, the hydrogens are sp^3 hybridized but stretching the C—H bonds leads to an increased crowding and bumping, and this is manifested by a steeper potential energy surface and an increase in k, the force constant.

Nitrogen–Hydrogen Stretching Frequencies

Much of what we discussed regarding C—H stretching frequencies is also applicable here. There are three major differences between the C—H and N—H stretching frequencies. First, the force constant for N—H stretching is stronger; there is a larger dipole moment associated with the N—H bond;

finally, the N—H bond is usually involved in hydrogen bonding. The stronger force constant leads to a higher frequency for absorption. The N—H stretching frequency is usually observed from 3500 to 3200 cm^{-1}. The larger dipole moment leads to a stronger absorption, and the presence of hydrogen bonding has a definite influence on the band shape and frequency position. The presence of hydrogen bonding has two major influences on spectra. First, it causes a shift towards lower frequency of all functional groups that are involved in hydrogen bonding; second, the peaks are generally broadened. Keep these two factors in mind as you examine the following spectra, regardless of what atoms and functional groups are involved in the hydrogen bonding.

The N—H stretching frequency is most frequently encountered in amines and amides. The following examples illustrate the behaviour of this functional group in a variety of circumstances.

Primary amines and amides derived from ammonia The N—H stretching frequency in primary amines and amides derived from ammonia has the same local symmetry as observed in CH_2. Two bands, a symmetric and an asymmetric stretch, are observed. It is not possible to assign the symmetric and asymmetric stretches by inspection, but their presence at approximately 3300 and 3340 cm^{-1} is suggestive of a primary amine or amide. These bands are generally broad, and a third peak at frequencies lower than 3300 cm^{-1}, presumably due to hydrogen bonding, is also observed. This is observed for *n*-butyl amine and benzamide.

Secondary amines and amides Secondary amines and amides show only one peak in the infrared. This peak is generally in the vicinity of 3300 cm^{-1}. The effect of hydrogen bonding on the breadth of the N—H peak could be seen in the spectrum.

Tertiary amines and amides Tertiary amines and amides from secondary amines have no observable N—H stretching band.

N–H bending motions We shall ignore most bending motions because these occur in the fingerprint region of the spectrum. One exception is the N–H bend that occurs at about 1600 cm^{-1}. This band is generally very broad and relatively weak. Since many other important bands occur in this region, it is important to note the occurrence of this absorption; lest it be mistakenly interpreted as another functional group. Most other functional groups absorbing in this region are either sharper or more intense.

Hydroxyl Stretch

Hydroxyl stretch is similar to the N—H stretch in that its hydrogen bonds more strongly. As a result, it is often broader than the N—H group. In those rare instances, when it is not possible to hydrogen bond, the stretch is found as a relative weak-to-moderate absorption at 3600–3650 cm^{-1}. In tri-t-butylmethanol, where steric hindrance prevents hydrogen bonding, a peak at 3600 cm^{-1} is observed. Similarly, for hexanol, phenol and hexanoic acid, gas phase and liquid phase spectra illustrate the effect of hydrogen bonding on both the O—H stretch and on the rest of the spectrum. It should be pointed out that, in general, while gas phase spectra are usually very similar, frequencies are generally shifted to slightly higher values in comparison to condensed phase spectra. Gas phase spectra that differ significantly from condensed phase spectra are usually taken as evidence for the presence of some sort of molecular association in the condensed phase.

The hydroxyl group in phenols and alcohols usually is found as a broad peak centered at about 3300 cm^{-1} in the condensed phase. The O—H of carboxylic acid is so strongly associated that the O—H absorption in these materials is often extended to approximately 2500 cm^{-1}. This extended absorption is clearly observed and serves to differentiate the O—H stretch of carboxylic acid from that of alcohol or phenol. In fact, carboxylic acids associate to form intermolecular hydrogen-bonded dimers both in the solid and liquid phases.

The Nitrile Group

The nitrile group is another reliable functional group that is generally easy to identify. There is a significant dipole moment associated with the C—N bond, which causes a significant change when it interacts with infrared radiation usually leading to an intense sharp peak at 2200–2280 cm^{-1}. Very few groups at this region absorb with this intensity. If another electronegative atom, such as halogen, is attached to the same carbon as the nitrile group, the intensity of this is markedly reduced.

The Carbon–Carbon Triple Bond

The C—C bond is not considered to be a very reliable functional group. This stems in part by considering that the reduced mass is likely to vary. However, it is characterized by a strong force constant, and because this stretching frequency falls in a region where very little else absorbs, 2100–2260 cm^{-1}, it can provide useful information. The terminal carbon triple bond (CC—H) is the most reliable and easiest to identify. We have earlier discussed the C—H stretching frequency coupled with a band at 3300 cm^{-1}; the presence of a band at approximately 2100 cm^{-1} is a strong indication of the —CC—H group.

An internal —C—C— is more difficult to identify and is often missed. Unless an electronegative atom, such as nitrogen or oxygen, is directly attached to the sp hybridized carbon, the dipole moment associated with this bond is small; stretching this bond would lead to a very small change. In cases where symmetry is involved, such as in diethyl acetylenedicarboxylate, there is no change in dipole moment, and this absorption peak is completely absent. In cases where this peak is observed, it is often weak and difficult to identify with a high degree of certainty.

The Carbonyl Group

The carbonyl group is probably the most ubiquitous group in organic chemistry. It comes in various disguises. Carbonyl is a polar functional group that frequently has the most intense peak in the spectrum. We shall discuss some of the typical acyclic aliphatic molecules that contain a carbonyl group. Subsequently, we shall deal with the effect of including carbonyl as a part of the ring. Finally, we shall comment on the effect of conjugation on carbonyl frequency.

Acyclic Aliphatic Carbonyl Groups

Esters, aldehydes and ketones Esters, aldehydes and ketones are common examples of molecules exhibiting a $C = O$ stretching frequency. The frequencies, 1735, 1725 and 1715 cm^{-1} respectively, are too close to make a clear distinction between them. However, aldehydes can be distinguished by examining the presence of both C—H (2750, 2850 cm^{-1}) and a carbonyl group.

Carboxylic acids, amides and carboxylic acid anhydrides Carboxylic acids, amides and carboxylic acid anhydrides round out the remaining carbonyl groups frequently found in aliphatic molecules. The carbonyl frequencies of these molecules, 1700–1730 cm^{-1} (carboxylic acid), 1640–1670 cm^{-1} (amide) and 1800–1830 cm^{-1}, 1740–1775 cm^{-1} (anhydride), allow for an easy differentiation when the following factors are taken into consideration.

A carboxylic acid can easily be distinguished from all the carbonyl-containing functional groups by noting that carbonyl, at 1700–1730 cm^{-1}, is strongly hydrogen-bonded and broadened as a result. In addition, it contains an O—H stretch that shows similar hydrogen-bonding as noted earlier.

Amides are distinguished by their characteristic frequency, which is the lowest carbonyl frequency observed for an uncharged molecule, 1640–1670 cm^{-1} (amide I). In addition, amides from ammonia and primary amines exhibit a weaker second band (amide II) at 1620–1650 cm^{-1} and 1550 cm^{-1}, respectively, when the spectra are run on the solids. Amides from secondary amines do not have a hydrogen attached at nitrogen and do not show an amide II band. The Amide I band is mainly attributed to the carbonyl stretch. Amide II involves several atoms, including the N—H bond. We shall return to the frequency of amide carbonyl when we discuss the importance of conjugation and the effect of resonance on carbonyl frequencies. The spectra of benzamide, a conjugated amide and N-methyl acetamide clearly identify the amide I and II bands. The spectrum of N,N-dimethyl acetamide illustrates an example of an amide from a secondary amine.

Anhydrides can be distinguished from other simple carbonyl-containing compounds in that they contain and exhibit two carbonyl frequencies. However, these frequencies are not characteristic of each carbonyl. Rather they are another example of the effects of local symmetry similar to what we have seen for the CH_2 and NH_2 groups. The motions involved here encompass the entire anhydride ($—(C=O—O—(O=C—)$) in a symmetric and asymmetric stretching motion of the two carbonyls. The two carbonyl frequencies often differ in intensity. It is possible neither to assign the peaks to the symmetric or asymmetric stretching motion by inspection nor to predict the more intense peak. However, the presence of two carbonyl frequencies and the magnitude of the higher frequency (1800 cm^{-1}) is a good indication of an anhydride.

Cyclic Aliphatic Carbonyl-containing Compounds

The effect on carbonyl frequency as a result of including a carbonyl group as part of a ring is usually attributed to ring strain. Generally, ring strain is believed to be relieved in large rings because of the variety of conformations available. However, as the size of the ring gets smaller, this option is not available and a noticeable effect is observed. The effect of increasing ring strain is to increase the carbonyl frequency, independent of whether the carbonyl is a ketone, part of a lactone (cyclic ester), anhydride or lactam (cyclic amide). The carbonyl frequencies for a series of cyclic compounds are summarized in Table 3.3.

Table 3.3 Effect of ring strain on the carbonyl frequencies of some cyclic molecules

Ring size	Ketone (cm^{-1})	Lactones (cm^{-1})	(Lactams cm^{-1})
3	cyclopropanone (1800)	–	–
4	cyclobutanone (1775)	β-propiolactone (1840)	–
5	Cyclopentanone (1751)	γ-butyrolactone (1750)	γ-butyrolactam (1690)
6	cyclohexanone (1715)	δ-valerolactone (1740)	δ-valerolactam (1668)
7	cycloheptanone (1702)	ε-caprolactone (1730)	ε-caprolactam (1658)

Carbon-Carbon double bond Unlike the $C—C$ bond, the $C=C$ bond stretch is not a very reliable functional group. However, it is characterized by a strong force constant, and because the effects of conjugation, which we will see, enhance the intensity of this stretching frequency, this absorption can provide useful and reliable information.

Terminal $C=CH_2$ In simple systems, the terminal carbon–carbon double bond ($C=CH_2$) is the most reliable and easiest to identify since the absorption is of moderate intensity at 1600–1675 cm^{-1}. We have earlier discussed the C—H stretching frequency of an sp^2 hybridized C—H. In addition, the terminal C $=$ CH$_2$ is also characterized by a strong band at approximately 900 cm^{-1}. Since this band falls in the fingerprint region, some caution should be exercised in its identification.

Internal C $=$ C An internal non-conjugated C $=$ C is difficult to identify and can be missed. The dipole moment associated with this bond is small; stretching this bond would lead to a very small change. In cases where symmetry is involved, such as in 4-octene, there is no change in dipole moment, and this absorption peak is completely absent. In cases where this peak is observed, it is often weak. In 2,5-dihydrofuran, it is difficult to assign the C $=$ C stretch because of the presence of other weak peaks in the vicinity. The band at approximately 1670 cm^{-1} may be the C $=$ C stretch. In 2,5-dimethoxy, 2,5-dihydrofuran, the assignment at 1630 cm^{-1} is easier but the band is weak.

There is one circumstance that may have a significant effect on the intensity of both internal and terminal olefins and acetylenes. The substitution of a heteroatom directly on the unsaturated carbon to produce, for example, a vinyl or acetylenic ether or amine may lead to a significant change in the polarity of the C $=$ C or C—C bond, and a substantial increase in intensity is observed. The C $=$ C in 2,3-dihydrofuran, is observed at 1617.5 cm^{-1} and is one of the most intense bands in the spectrum. Moving the C $=$ C bond over one carbon gives 2,5-dihydrofuran, attenuates the effect and results in a weak absorption.

Aromatic Ring Breathing Motions

Benzene rings are frequently encountered in organic chemistry. Although we may write benzene as a six-membered ring with three double bonds, this is not a good representation of the structure of the molecule. The vibrational motions of a benzene ring are not isolated but involve the entire molecule. To describe one of the fundamental motions of benzene, consider some imaginary lines passing through the centre of the molecule and extending out through each carbon atom and beyond. A symmetric stretching and compression of all the carbon atoms of benzene along each line is an example of the ring breathing motion. Simultaneous expansions and compressions of these six carbon atoms lead to other ring-breathing motions. These vibrations are usually observed between 1450 and 1600 cm^{-1} and often lead to four observable absorptions of variable intensity. As a result of symmetry, benzene does not exhibit these bands; however, most benzene derivatives do. Usually two or three of these bands are sufficiently separate from other absorptions that they can be identified with a reasonable degree of confidence. The least reliable of these bands are those observed at approximately 1450 cm^{-1} where C—H bending motions are observed. Since all organic molecules that contain hydrogen are likely to have a C—H bond, absorptions observed at 1450 cm^{-1} are not very meaningful and should usually be ignored. Two of the four bands around 1600 cm⁻¹ are observed in *ortho* and *meta* xylene, identified by the Greek letter, and a third band at about 1500 cm⁻¹ is assigned.

Nitro Group

The final functional group we are going to discuss is the nitro group. In addition to being an important functional group in organic chemistry, it also initiates our discussion of the importance of using resonance to predict effects in infrared spectroscopy. A Kekule or Lewis structure for the nitro group is shown in Figure 3.8. You will find that it would be necessary to involve all five valence electrons of nitrogen and use them to form the requisite number of bonds to oxygen. This will lead to a positive charge on nitrogen and a negative charge on one oxygen. As a result of resonance, we will delocalize the negative charge on both oxygens, and this leads to an identical structure as shown in Figure 3.8. Since the structures are identical, we would expect the correct structure to be a resonance hybrid of the two. In terms of geometry, we would expect the structure to be a static average of the two geometric structures both in terms of bond distances and bond angles. Based on what we observed for the CH_2 and NH_2 stretch, we would expect a symmetric and an asymmetric stretch for the N—O bond in the nitro group halfway betweenthe N=O and N—O stretches. Two strong bands are observed, one at 1500–1600 cm⁻¹ and a second between 1300 and 1390 m⁻¹.

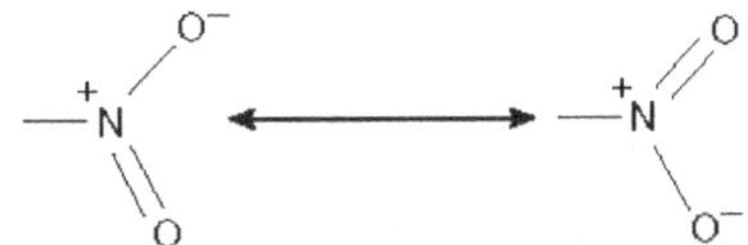

Figure 3.8   Resonance hybrid of nitro group

Effect of Resonance and Conjugation on Infrared Frequencies

The carboxylate anion is represented as a resonance hybrid (Figure 39).

Figure 3.9   Resonance hybrid of carboxylate anion

Unlike the nitro group which contained functional groups, the carboxyl group is made up of a resonance hybrid between a carbon–oxygen single bond and a carbon–oxygen double bond. According to resonance, we would expect the C—O bond to be an average between a single and double bond or approximately equal to a bond and a half. We can use the carbonyl frequency of an ester of 1735 cm^{-1} to describe the force constant of the double bond. We have not discussed the stretching frequency of a C—O single bond for the simple reason that it is quite variable and that it falls in the fingerprint region. However, the band is known to vary from 1000 to 1400 cm^{-1}. For the purpose of this discussion, we will use an average value of 1200 cm^{-1}. The carbonyl frequency for a bond and a half would be expected to fall halfway between 1735 and 1200 or at approximately 1465 cm^{-1}. The carboxyl group has the same symmetry as the nitro and CH$_2$ groups. Both symmetric and asymmetric stretch should be observed. The infrared spectrum of sodium benzoate should show asymmetric and symmetric stretch at 1410 and 1560 cm^{-1}. The peak observed averages to 1480 cm^{-1}, and it is in good agreement with the average frequency predicted for a carbon–oxygen bond with a bond order of 1.5. While this is a qualitative argument, it is important to realize that the carboxylate anion does not show the normal carbonyl and normal C—O single bond stretches (at approximately 1700 and 1200 cm^{-1}) suggested by each of the static structures above.

In the cases of the nitro group and the carboxylate anion, both resonance forms contribute equally to describing the ground state of the molecule.

Carbonyl Frequencies

Most carbonyl stretching frequencies are found at approximately 1700 cm^{-1}. A notable exception is the amide carbonyl which is observed at approximately 1600 cm^{-1}. This suggests that the following resonance form makes a significant contribution to describing the ground state of amides (Figure 3.10).

Figure 3.10 Resonance hybrid of amides

Note that resonance forms that lead to charge separation are not considered to be very important. However, the following information supports the importance of resonance in amides. X-ray crystal structures of amides show that the amide functional group is planar in the solid state. This suggests sp^2 hybridization at nitrogen rather than sp^3. In addition, the barrier to rotation about the carbon–nitrogen bond has been measured. Unlike the barrier of rotation of most aliphatic C—N bonds, which are of the order of a few kcal/mol, the barrier to rotation about the carbon–nitrogen bond in dimethyl formamide is approximately 18 kcal/mol. This suggests an important contribution of the dipolar structure to the ground state of the molecule, and the observed frequency of 1600 cm⁻¹, according to the arguments given above for the carboxylate anion, is consistent with more C—O single bond character than would be expected otherwise.

The conjugation of a carbonyl wih a C=C bond is thought to cause an increase in resonance interaction. Again the resonance forms lead to charge separation, which clearly de-emphasizes their importance.

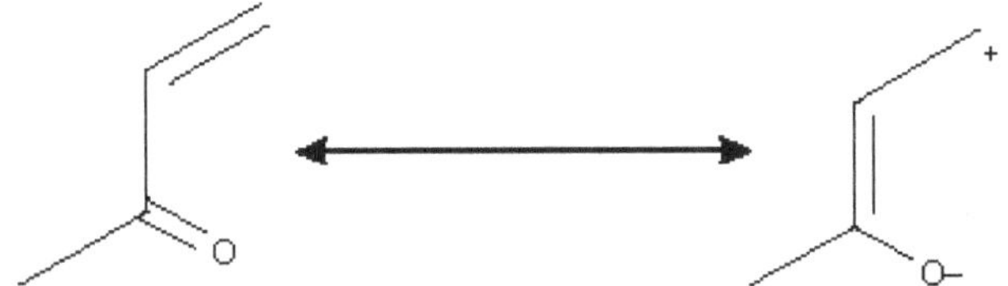

Figure 3.11   Resonance forms of exone

However, this conjugative interaction is useful in interpreting several features of the spectrum. First, it predicts the small but consistent shift of approximately 10 cm⁻¹ to lower frequency, observed when carbonyls are conjugated to double bonds or aromatic rings. Next, the dipolar resonance form suggests a more polar C=C than that predicted for an unconjugated C=C. In terms of the change in dipole moment, contributions from this structure suggest that the intensity of infrared absorption of a C=C double bond would increase relatively to an unconjugated system. Conjugation is associated with an increase in intensity ofthe C=C stretching frequency. However, not all conjugated carbonyls appear as multiplets. The resolution of this additional complicating feature can be achieved if we consider that conjugation requires a fixed conformation. For most conjugated carbonyls, two or more conformations are possible. The *s-cis* form is shown in (Figure 3.11) above and the *s-trans* form is shown in (Figure 3.2).

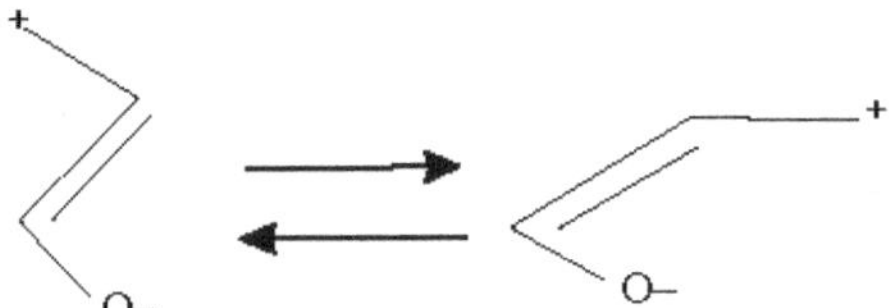

Figure 3.12   Conformations of conjugate carboxyl

If the resonance interactions in these two forms differ, the effect of resonance on the carbonyl will differ, leading to similar but different frequencies. The presence of multiple carbonyl frequencies is a good indication of a conjugated carbonyl. In some conjugated systems such as benzaldehyde, different types of bonds have characteristic regions of the spectrum where they absorb as shown in Figure 3.13.

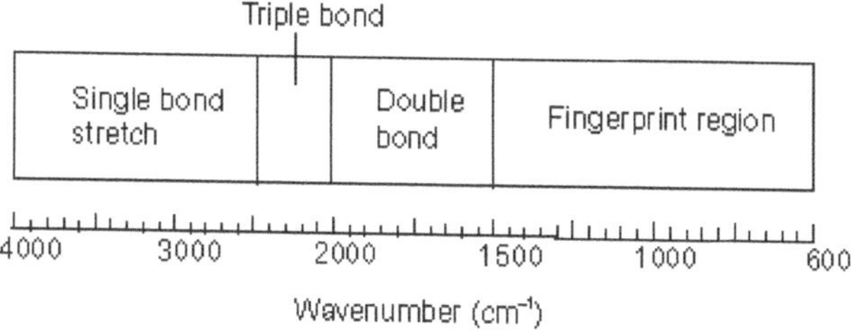

Figure 3.13   Different regions of IR spectra

Most functional groups absorb above 1500 cm⁻¹. The region below 1500 cm⁻¹ is known as the fingerprint region. Each and every molecule produces a unique pattern here; so if an unknown sample produces a spectrum which matches that of a known compound, the sample can be confirmed to be that compound.

Applications of Infrared Spectroscopy

Identification of an organic compound   The infrared spectrum of an organic compound, particularly in the fingerprint region, is very useful in establishing the identity of the compound. The molecules containing the same functional group may have similar absorptions above 1500 cm⁻¹ but their spectra differ significantly in the fingerprint region.

Structure determination   Infrared spectroscopy is used to establish the structure of a compound. For example, the infrared spectra of amino acids exhibit bands for ionized carboxylic acids and amine salts but none for free amino and carboxylic group. Thus, amino acids exist as zwitterions.

$$H_3N^+ - CH - COO^-$$

Detection of impurities   The use of infrared spectroscopy for the detection of impurities may sometimes be misleading because infrared bands are not often very intense and, therefore, small amounts of impurities may remain undetected. However, the presence of impurities can be detected provided they absorb fairly strongly in that part of the spectrum where the main component does not.

Detection of functional groups (qualitative analysis)   It is possible to establish a co-relationship between infrared absorptions and functional groups. For example, the oxygen in a compound can be present only as O—H, C=O or C—O—C. The presence or absence of absorption in the carbonyl region (1870–1650 cm⁻¹) or hydroxyl region (3700–3200 cm⁻¹) can serve to ascertain some of these possibilities.

Quantitative analysis   Infrared spectroscopy has been employed for the quantitative estimation of an organic mixture. It is possible to estimate the ratio of components of a mixture by (i) measuring the relative intensities of absorption bands characteristic of each component, (ii) the knowledge of the absorbance of the pure compound.

Kinetic study of reactions   Infrared spectroscopy is often used to follow the progress of a chemical reaction. Since most chemical reactions involve changes in the functional groups of the reacting moieties, periodic examination of only a small portion of the spectrum may be sufficient to reveal the progress of a chemical reaction. For example, the oxidation of alcohol to aldehyde or ketone or the reduction of ketone to alcohol can be followed by the appearance or disappearance of the carbonyl bond.

Distinction between inter- and intramolecular hydrogen bonding   Hydrogen bonding can be classified into two types—inter- and intramolecular. Intermolecular hydrogen bonding may be between two molecules or between the molecule and the solvent. The former can be eliminated by using very dilute solutions and the latter by aprotic solvents. Intramolecular hydrogen bonding is independent of solvent concentration. The free O—H bond of alcohols is at 3600 cm^{-1}, and hydrogen-bonded —OH band is broad and at 3550–3200 cm^{-1}. Thus, in intermolecular hydrogen-bonded compounds, when diluted, the low-frequency, broad intense O—H (bonded) band diminishes and the high frequency, sharp free O—H band of unbonded hydroxyl group appears.

Distinction between geometrical isomers   In general, *trans*-isomers are more symmetric than *cis*, and consequently the vibrations of the former result in little or no change in dipole moment. Hence, *trans*-isomers give a simpler IR spectrum than *cis*-isomers. If the samples of both isomers of a compound are available, then the *cis*- and *trans*-isomers can be identified.

SPECTRAL INTERPRETATION

A sample with chemical formula $(C_3H_6O_2)$ gives the following spectrum:

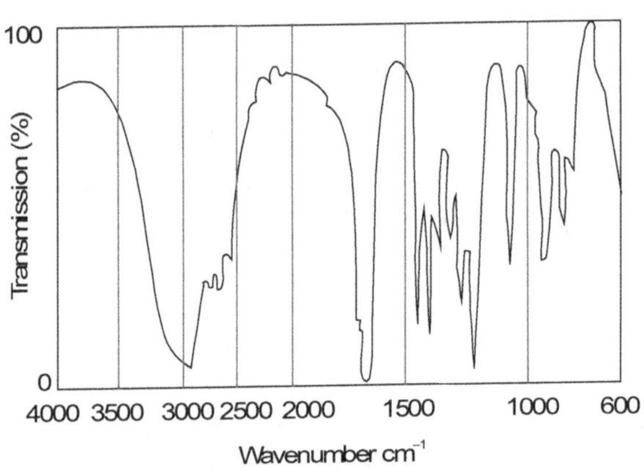

Figure 3.14   Spectrum of sample $(C_3H_6O_2)$

This shows a strong absorption at just over 1710 cm^{-1} and a medium absorption at 2500–3300 cm^{-1}. From Table 3.1, it is known that these bonds correspond to the CO and O—H groups found in a carboxylic acid. This information of the chemical formula tells us that the structure must be propanoicacid:

Propanoic acid

Once the molecular formula of an unknown is known, it is simple to determine the degree of unsaturation. The degree of unsaturation is the sum of the number of carbon—carbon multiple bonds and rings. Each reduces the number of hydrogens (or any other element with a valence of one) by two. There is a general formula that can be used:

$$\text{Unsaturation number, } U = \#\ C + 1 - 1/2\ (X - N)$$

where, X = monovalent atoms

N = trivalent atoms, and

C = tetravalent atoms. Note that divalent atoms are not counted.

Consider C_6Cl_6 as an example.

Degree of unsaturation, $U = 6 + 1 - 1/2\ (6 - 0)$

$$= 7 - 3 = 4$$

An unsaturation factor of four is required for a single benzene ring.

Consider C_5H_9NO as another example.

Degree of unsaturation, $U = 5 + 1 - 1/2\ (9 - 1)$

$$= 2$$

Clearly, this compound cannot have a benzene ring but the oxygen atom could be part of a carbonyl group.

The application of degree of unsaturation to the interpretation of an infrared spectrum is quite straightforward. Clearly, some functional groups can be eliminated by composition. Amines, amides, nitriles and nitro groups can be eliminated if the molecule does not contain any nitrogen. Alternatively, everything but amines can be eliminated if the molecular formula contains nitrogen and no degrees of unsaturation.

The following steps prove to be useful regardless of the structure of the unknown.

1. Look in the carbonyl region, typically 1800–1620 cm^{-1}, for a strong absorption band. So many classes of compounds contain a carbonyl group

(carboxylic acids, esters, amides, ketones, aldehydes, etc.), and absorption is so obvious that it is the perfect starting point. If there is no carbonyl absorption, you have eliminated a large number of possibilities.

2. Examine the C—H stretching frequencies at 3000 cm^{-1}. Absorption bands at frequencies slightly larger than 3000 cm^{-1} are indicative of vinyl or/and aromatic hydrogens. The presence of these peaks should be consistent with the degree of unsaturation of your molecule. The absence of absorption above 3000 cm^{-1} but the presence of some unsaturation in the molecular formula are consistent with a cyclic compound.

   If your degree of unsaturation is 4 or greater, look for 2 to 4 absorption peaks between 1600 and 1450 cm^{-1} and weak peaks at 2000–1667 cm^{-1}. These are characteristic of aromatic compounds.

3. Next, look for a doublet at 2750 and 2850 cm^{-1} characteristic of an aldehyde. The presence of these two bands should also be accompanied by a strong absorption at approximately 1700 cm^{-1}. Most spectra display strong absorption in the 1800–1700 cm^{-1} region. If your spectrum does, check to see if the carbonyl is a closely spaced doublet or multiplet. Closely spaced multiplicity in the carbonyl region accompanied by C—H absorption at 3000–3100 cm^{-1} is a frequent characteristic of an α,β-unsaturated carbonyl compound. Check to make sure that the carbonyl frequency is consistent with conjugation.

4. If your unknown contains broad absorption from 3600–3000 cm^{-1}, your molecule could have an O—H or N—H stretch. Check the multiplicity of this peak. A doublet is characteristic of a primary amine and/or amide derived from ammonia. Check the carbonyl region at around 1650–1600 cm^{-1}. Two bands in this region are consistent of an amide from ammonia or a primary amine. Remember that a broad and relatively weak band at about 1600 cm^{-1} is characteristic of N—H bending. Usually, you will see this band only in amines, since the carbonyl group of the amide will interfere. Be sure to look for the effect of hydrogen bonding that usually results in a general broadening of the groups involved.

5. If the broad band starting at 3600 cm^{-1} expands to nearly 2400 cm^{-1}, look for the presence of a broad carbonyl at approximately 1700 cm^{-1}. This extremely broad OH band is only observed in carboxylic acids and enols from β-diketones. The presence of a relatively intense but broad band at approximately 1700 cm^{-1} is a good evidence for carboxylic acid.

Fourier transform Infrared Spectrometer

Fourier transform infrared (FTIR) spectrometer is a single beam instrument. Unlike a double beam instrument that simultaneously corrects for absorptions due to atmospheric water vapour and carbon dioxide, most FTIR spectrometers correct for the background absorption by storing an interferogram and background spectrum before recording the spectrum. An interferogram contains the same information as a regular spectrum, i.e., frequency vs. intensity, but this information is stored in the form of intensity vs. time.

The Fourier transform is the mathematical process that converts the information from intensity and time to intensity and frequency.

One of the major advantages of an FTIR spectrometer is that it takes in the order of a second to record an entire spectrum. This makes it very convenient to record the spectrum a number of times and display an average spectrum. Since the signal to noise ratio varies as the square root of the number of scans averaged, it is easy to obtain a good signal to noise ratio on this type of instrument, even if you have very little sample. On a typical spectrum, you should average at least four spectra.

Fourier transform spectrometers have recently replaced dispersive instruments due to their superior speed and sensitivity. They have greatly extended the capabilities of infrared spectroscopy and have been applied to many areas that are very difficult or nearly impossible to analyse with dispersive instruments. Instead of viewing component frequencies sequentially as with a dispersive IR spectrometer, all frequencies can be examined simultaneously with FTIR spectroscopy.

Components of FTIR Spectrometer

There are three basic components in an FTIR spectrometer: radiation source, interferometer and detector. A simplified optical layout of a typical FTIR spectrometer is shown in Figure 3.15. The same kinds of radiation sources are used for both dispersive and Fourier transform spectrometers. However, the source is more often water-cooled in FTIR instruments to provide better power and stability.

In contrast, a completely different approach is taken in an FTIR spectrometer to differentiate and measure the absorption at component frequencies. The monochromator is replaced by an interferometer, which divides radiant beams, generates an optical path difference between the beams, and then recombines them in order to produce repetitive interference signals measured as a function of optical path difference by a detector. The interferometer produces interference signals, which contain infrared spectral information generated after passing through a sample.

The most commonly used interferometer is a Michelson interferometer. It consists of three active components: a moving mirror, a fixed mirror and a beamsplitter (Figure 3.15). The two mirrors are perpendicular to each other. The beamsplitter is a semi-reflecting device and is often made by depositing a thin film of germanium onto a flat KBr substrate. The radiation from the broadband IR source is collimated and directed into the interferometer, which impinges on the beamsplitter. At the beamsplitter, half the IR beam is transmitted to the fixed mirror and the remaining half is reflected to the moving mirror. After the beams are reflected from the two mirrors, they are recombined at the beamsplitter. Due to changes in the relative position of the moving mirror to the fixed mirror, an interference pattern is generated. The resulting beam then passes through the sample and is eventually focused on the detector.

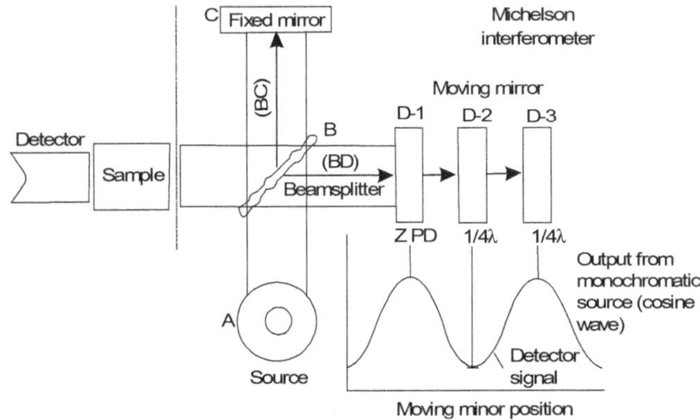

Figure 3.15   A typical FTIR spectrometer

The detector response for a single-frequency component from the IR source is first considered. This simulates an idealized situation where the source is monochromatic, such as a laser source. As previously described, the differences in the optical paths between the two split beams are created by varying the relative position of the moving mirror to the fixed mirror. If the two arms of the interferometer are of equal length, the two split beams travel through the same path length. The two beams are totally in phase with each other; thus, they interfere constructively and lead to a maximum detector response. This position of the moving mirror is called the point of zero path difference (ZPD). When the moving mirror travels in either direction by the distance $\lambda/4$, the optical path (beamsplitter–mirror–beamsplitter) is changed by 2 ($\lambda/4$), or $\lambda/2$. The two beams are 180° out of phase with each other and thus interfere destructively. As the moving mirror travels another $\lambda/4$, the optical path difference is now $2(\lambda/2)$, or λ. The two beams are again in phase with each other and result in another constructive interference.

When the mirror is moved at a constant velocity, the intensity of radiation reaching the detector varies in a sinusoidal manner to produce the interferogram output as shown in Figure 3.15. The interferogram is the record of the interference signal. It is actually a time domain spectrum and records the detector response changes versus time within the mirror scan. If the sample happens to absorb at this frequency, the amplitude of the sinusoidal wave is reduced by an amount proportional to the amount of sample in the beam.

The extension of this procedure to three component frequencies would result in a more complex interferogram, which is the summation of three individual modulated waves. In contrast to this simple, symmetric interferogram, the interferogram produced with a broadband IR source displays extensive interference patterns. It is a complex summation of superimposed sinusoidal waves, each wave corresponding to a single frequency. When this IR beam is directed through the sample, the amplitudes of a set of waves are reduced by absorption if the frequency of this set of waves is the same as one of the characteristic frequencies of the sample.

The interferogram contains information over the entire IR region to which the detector is responsive.

A mathematical operation known as Fourier transformation converts the interferogram (a time domain spectrum, displaying intensity versus time within the mirror scan) to the final IR spectrum, which is the familiar frequency domain spectrum showing intensity versus frequency. This also explains how the term Fourier transform infrared spectrometry is created.

The detector signal is sampled at small, precise intervals during the mirror scan. The sampling rate is controlled by an internal, monochromatic beam from a helium–neon (He–Ne) laser focused on a separate detector.

The two most popular detectors for a FTIR spectrometer are deuterated triglycine sulphate (DTGS) and mercury cadmium telluride (MCT). The response times of many detectors (for example, thermocouple and thermistor) used in dispersive IR instruments are too slow for the rapid scan times (one second or less) of the interferometer. The DTGS detector is a pyroelectric detector that delivers rapid responses because it measures the changes in temperature rather than the value of temperature. The MCT detector is a photon (or quantum) detector that depends on the quantum nature of radiation and also exhibits very fast responses. While DTGS detectors operate at room temperature, MCT detectors must be maintained at liquid nitrogen temperature (77 K) to be effective. In general, the MCT detector is faster and more sensitive than the DTGS detector.

Problems and Solutions

1. Calculate the vibrational degrees of freedom in the following molecules:

    i. O_2 ii. N_2O iii. CH_2O

    iv. CH_4 v. C_6H_6 vi. $CHCl_3$

    For linear molecules, the vibrational degrees of freedom = $3n - 5$. For non-linear molecules, the vibrational degrees of freedom = $3n - 6$. Hence, the vibrational degrees of freedom are (i) one (ii) three (iii) six (iv) nine (v) thirty (vi) six.

2.   Ethylene and ethyne, unlike propene and propyne, have no carbon-to-carbon multiple bond stretching bands. Explain.

Ethyne and ethylene are symmetrical molecules and do not give rise to a change in dipole moment when excited. A change in dipole moment occurs when propene and propyne are excited and a band is observed.

3.   Calculate wavenumber of radiation required to vibrate O–H bond, if the force constant is 7.7×10^5.

$$v = \frac{1}{a}\sqrt{\frac{k}{\mu}} = \frac{1}{2\pi}\sqrt{\frac{7.7\times10^5}{0.941\times1.67\times10^{-24}}}$$

$$= 1.11\times10^{14}\,\text{Hz}$$

$$\overline{v} = \frac{v}{c} = \frac{1.11\times10^{14}}{3\times10^{10}}\,\text{cm}^{-1} = 3700\,\text{cm}^{-1}$$

REVIEW QUESTIONS

1.   Describe fundamental vibrations and overtones.

2.   Calculate the vibrational degrees of freedom of

   i. O_2 ii. CH_4 iii. C_6H_6

(Ans: i. one ii. nine iii. thirty)

3.   Distinguish between the following pairs of compounds using IR.

   i.   *cis* and *trans* stilbene   ii. CH_3CH_2CHO and $CH_3\overset{\displaystyle O}{\underset{\displaystyle C}{\|}}CH_3$

   iii. $(CH_3)_3\,N$ and $CH_3CH_2CH_2NH_2$

4.   Arrange the following bonds in order of their decreasing vibrational frequencies.

   C—Cl, C—Br, C—C, C—H, C—O, C—I

5.   What materials are used for making prisms for use in IR spectroscopy?

6.   How can primary, secondary and tertiary alcohols be distinguished using IR?

7.   What types of detectors are used in IR?

8.   Give any three applications of IR spectroscopy.

9.   Maleic acid absorbs at a higher frequency compared to fumaric acid. Account for this.

10.   Citing examples, discuss the inductive and mesomeric effects influencing the carbonyl absorption frequency.

11. Write a note on fingerprint region.

12. Describe briefly the KBr pellet technique.

13. Which of the following molecules will show infrared spectrum and why?

    $HCl, N_2O, H_2, CO_2, CH_4$

    (Ans: All except H_2 as it is a homonuclear molecule)

Chapter 4

NUCLEAR MAGNETIC
RESONANCE SPECTROSCOPY

Introduction

Radio waves are the lowest form of electromagnetic radiation that is useful in analytical chemistry. The quantity of energy involved in radio frequency (RF) radiation is very small. It is too small to vibrate, rotate or excite an atom or a molecule electronically. However, it is sufficient to affect the nuclear spin of the atoms that make up the molecule. As a result, the spinning nuclei of atoms in a molecule can absorb RF radiation and change the direction of the spinning axis. However, this phenomenon requires the presence of an external magnetic field. This is the principle of **nuclear magnetic resonance spectroscopy**.

The phenomenon of nuclear resonance was first observed in 1946 by Purcell, Torrey and Pound at Harvard and then by Bloch, Hansen and Packard at Stanford, who shared the Nobel Prize for this discovery. The first systematic applications to organic chemistry were reported in 1953 by Meyer Saika and Gutowsky.

Spin active nuclei: energy absorption and relaxation

Spin is a fundamental property of nature, e.g. electrical charge or mass. Spin comes in multiples of 1/2 and may be positive or negative. Individual unpaired electrons, protons and neutrons possess a spin of 1/2. In the deuterium atom (^{2}H) with one unpaired electron, one unpaired proton and one unpaired neutron, the total electronic spin is 1/2 and the total nuclear spin is 1.

Two or more particles with spins having opposite signs can pair up to eliminate the observable manifestation of spin. An example is helium. In nuclear magnetic resonance (NMR), unpaired nuclear spins are of greater importance.

The nuclei of many elemental isotopes have a characteristic spin (I). Some nuclei have integral spins (e.g. $I = 1, 2, 3...$); some have fractional spins (e.g. $I = 1/2, 3/2, 5/2...$); a few have no spin, i.e., $I = 0$ (e.g. ^{12}C, ^{16}O, ^{32}S,...). Isotopes of particular interest and use to organic chemists are ^{1}H, ^{13}C, ^{19}F and ^{31}P, all of which have I = 1/2. Some of the nuclei routinely used in NMR are shown in Table 4.1.

Table 4.1 Nuclei commonly used in NMR

Nuclei	Unpaired protons	Unpaired neutrons	Net spin	γ(MHz/T)
^{1}H	1	0	1/2	42.58
^{2}H	1	1	1	6.54
^{31}P	1	0	1/2	17.25
^{23}Na	1	2	3/2	11.27
^{14}N	1	1	1	3.08
^{13}C	0	1	1/2	10.71
^{19}F	1	0	1/2	40.08

Since the analysis of 1/2 spin state is straightforward, our discussion of NMR will be limited to these and other I = 1/2 nuclei.

When a proton is placed in an external magnetic field, the spin vector of the particle aligns itself with the external field, just like how a magnet would. There is a low-energy configuration or state where the poles are aligned N–S–N–S and a high-energy state N–N–S–S.

Transitions

The above proton can undergo a transition between the two energy states by the absorption of a photon. A proton in the lower energy state absorbs a photon and ends up in the upper energy state. The energy of this photon must exactly match the energy difference between the two states. The energy, E, of a photon is related to its frequency, ν, Planck's constant (h = 6.626 × 10^{-34} J s).

$$E = h\nu$$

In NMR and MRI, the quantity ν is called resonance frequency and Larmor frequency respectively.

Calculating Transition Energy

The nucleus generates a small magnetic field as a result of spinning. The nucleus, therefore, possesses a magnetic moment, m, which is proportional to its spin I.

$$\mu = \frac{\gamma I h}{2\pi}$$

The constant, γ is called the **magnetogyric ratio** and is a fundamental nuclear constant, which has a different value for every nucleus; and h is the Planck's constant.

The energy of a particular energy level is given by

$$E = -\frac{\gamma h}{2\pi} mB$$

where, B is the strength of the magnetic field at the nucleus.

The difference in energy between the levels (the transition energy) can be found from

$$\Delta E = \frac{\gamma h B}{2\pi}$$

This means that if the magnetic field, B, is increased, so is ΔE. It also means that if a nucleus has a relatively large magnetogyric ratio, then ΔE is correspondingly large.

In the NMR experiment, the frequency of the photon is in the RF range. In NMR spectroscopy, n is between 60 and 800 MHz for hydrogen nuclei. In clinical MRI, n is typically between 15 and 80 MHz for hydrogen imaging.

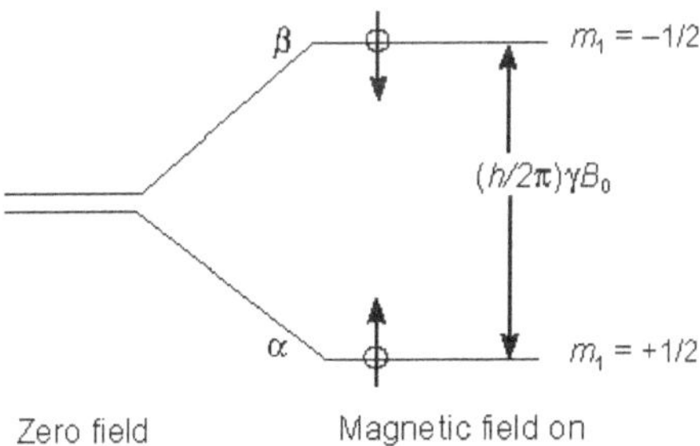

Figure 4.1   Two-spin states of nucleus in magnetic field

For the nuclei of spin 1/2, the energy difference between the two spin states at a given magnetic field strength will be proportional to their magnetic moments (Figure 4.1). For the four common nuclei (^{1}H, ^{19}F, ^{31}P and ^{13}C), the magnetic moments μ are 2.7927, 2.6273, 1.1305 and 0.7022 respectively. Figure 4.2 gives the approximate frequencies that correspond to the spin-state energy separations for each of these nuclei in an external magnetic field of 2.34 T. The formula (which is shaded grey shows the direct correlation of frequency (energy difference) with magnetic moment (h = Planck's constant).

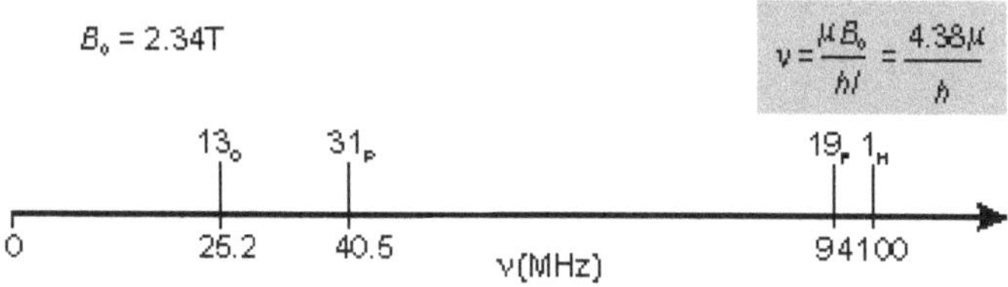

Figure 4.2 Spin-state energy separations of nuclei

When energy is absorbed by the nucleus, the angle of precession, θ, will change (Figure 4.3). For a nucleus of spin 1/2, the absorption of radiation 'flips' the magnetic moment so that it opposes the applied field (the higher energy state).

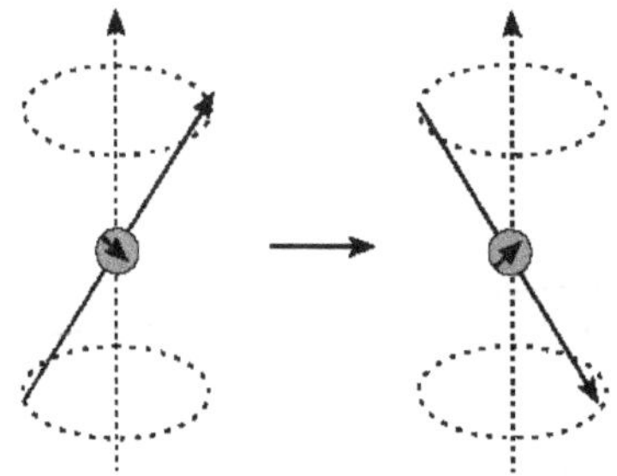

Figure 4.3 Precession of nucleus with the field and against the field

It is important to realize that only a small proportion of "target" nuclei are in the lower energy state (and can absorb radiation). There is a possibility that, by exciting these nuclei, the populations of the higher and lower energy levels will become equal. If this occurs, then there will be no further absorption of radiation. The spin system is **saturated**. The possibility of saturation means that we must be aware of the relaxation processes, which can return the nuclei to the lower energy state.

Relaxation Processes

Ideally, an NMR spectroscopist would like relaxation rates to be fast, but not too fast. If the relaxation rate is fast, then saturation is reduced. If the relaxation rate is too fast, line-broadening in the resultant NMR spectrum is observed. There are two major relaxation processes:

- Spin–lattice (longeitudinal) relaxation
- Spin–spin (transverse) relaxation

Spin-lattice relaxation   In an NMR experiment, the sample in which the nuclei are held is called the **lattice**. Nuclei in the lattice are in vibrational and rotational motion, which creates a complex magnetic field. The magnetic field caused by the motion of nuclei within the lattice is called the **lattice field**. Lattice field has many components. Some of these components will be equal in frequency and phase to the Larmor frequency of the nuclei of interest. These components of the lattice field can interact with nuclei in the higher energy state, and cause them to lose energy (returning to the lower state). The energy that a nucleus loses increases the amount of vibration and rotation within the lattice, resulting in a tiny rise in the temperature of the sample.

The relaxation time, T_1 (the average lifetime of nuclei in the higher energy state), depends on the magnetogyric ratio of the nucleus and the mobility of the lattice. As mobility increases, the vibrational and rotational frequencies increase, making it more likely for a component of the lattice field to be able to interact with excited nuclei. However, at extremely high mobilities, the probability of a component of the lattice field being able to interact with excited nuclei decreases.

Spin-spin relaxation   Spin–spin relaxation describes the interaction between neighbouring nuclei with identical precessional frequencies but differing magnetic quantum states. In this situation, the nuclei can exchange quantum states; a nucleus in the lower energy level will be excited, while the excited nucleus relaxes to the lower energy state. There is no **net** change in the populations of the energy states, but the average lifetime of a nucleus in the excited state will decrease. This can result in line-broadening.

Spectra and molecular structure: Elucidation of Proton NMR Spectra

A simple NMR spectrum provides three kinds of information.

1.   Chemical shift, which identifies the types of protons based on their electronic environment.
2.   Spin–spin splitting patterns, which help to identify neighbouring protons.
3.   Peak areas or intensities, which are proportional to the number of protons causing a given resonance line.

Chemical Shift

Unlike infrared and UV–visible spectroscopy, where absorption peaks are uniquely located by a frequency or wavelength, the location of different NMR resonance signals is dependent on both the external magnetic field strength and the RF. Since no two magnets can have exactly the same field, resonance frequencies will vary accordingly, and an alternative method for characterizing and specifying the location of NMR signals is needed. This problem can be illustrated by eleven different compounds shown in Figure 4.4. Although the eleven resonance signals are distinct and well-separated, an unambiguous numerical locator cannot be directly assigned to each.

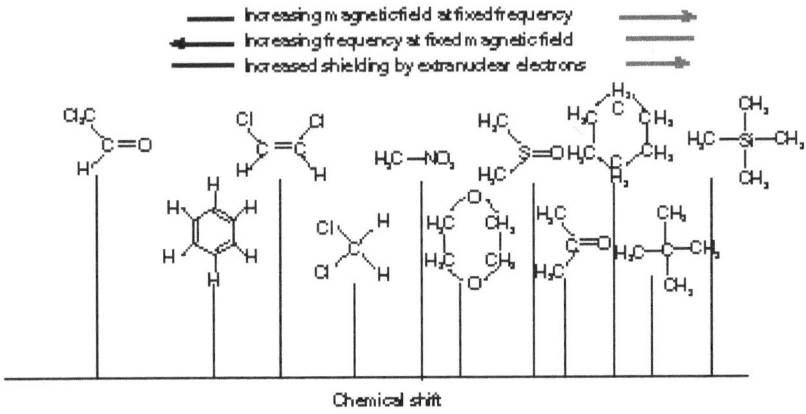

Increasing magnetic field at fixed frequency

Increasing frequency at fixed magnetic field

Increased shielding by extranuclear electrons

Chemical shift

Figure 4.4   ¹H NMR resonance signals for some compounds

One method of solving this problem is to report the location of an NMR signal in a spectrum relative to a reference signal from a standard compound added to the sample. Such a reference standard should be chemically non-reactive and should be easily removed from the sample after the measurement. Also, it should give a single sharp NMR signal that does not interfere with the resonances normally observed for organic compounds. **Tetramethylsilane** $[(CH_3)_4Si]$, usually referred to as **TMS**, meets all these characteristics, and has become the reference compound of choice for proton and carbon NMR. Since the separation or dispersion of NMR signals depends on magnetic field, one additional step must be taken in order to provide an unambiguous location unit. To correct these frequency differences for their field dependence, we divide them by the spectrometer frequency (100 or 500 MHz in the example). The resulting number would be very small, since we divided Hz by MHz, and so we multiply the result by a million (10^6). This operation gives a locator number called the **chemical shift**, having units of parts per million (ppm) and designated by the symbol δ.

Chemical shifts can be calculated using the formula:

$$\delta = \frac{v - v_0}{v_0} \times 10^6$$

where, v and v_0 are the frequency of the unknown and reference (TMS), respectively.

As mentioned earlier, the amount of shielding, and therefore the frequency separation between two lines, will increase with increasing field. However, the chemical shifts calculated with the above equation are field- independent. Thus, shifts measured at 100 MHz can be compared directly with those obtained at 500 MHz. The chemical shifts of different types of protons are provided in Table 4.2.

Table 4.2   Characteristic nuclear magnetic spectroscopy—Proton (H) chemical shifts

Type of proton	Structure	Chemical shift, ppm
Cyclopropane	C_3H_6	0.2
Alkane—primary H	$R—CH_3$	0.9
Alkane—secondary H	$R_2—CH_2$	1.3
Alkane—tertiary H	$R_3—C—H$	1.5
Alkene		4.6–5.9
Alkyne		2–3
Aromatic (arene)	Ar—H	6–8.5
Aromatic—aliphatic on ring	Ar—C—H	2.2–3
		1.7
Fluorides	H—C—F	4–4.5
Chlorides	H—C—Cl	3–4
Bromides	H—C—Br	2.5–4
Iodides	H—C—I	2–4
Alcohols	H—C—OH	3.4–4
Ethers	H—C—OR	3.3–4
Esters—H on 1st C of ...oate	RCOO—C—H	3.7–4.1
Esters—H on 1st C of ...ic	H—C—COOR	2–2.2
Methanoic acid	H—C—COOH	2–2.6
		2–2.7
Aldehyde		9–10
Hydroxylic	R—H—OH	1–5.5
Phenolic	Ar—OH	4–12
Enolic		15–17
Carboxylic	RCOOH	10.5–12
Amino	$R—NH_2$	1–5

In order to take the NMR spectra of a solid, it is usually necessary to dissolve the solid in a suitable solvent. Early studies used carbon tetrachloride for this purpose, since it has no hydrogen that could introduce an interfering signal. Unfortunately, CCl_4 is a poor solvent for many polar compounds and is also toxic. Deuterium labelled compounds such as deuterium oxide (D_2O), chloroform-d ($DCCl_3$), benzene-d_6 (C_6D_6), acetone-d_6 (CD_3COCD_3) and DMSO-d_6 (CD_3SOCD_3), are now widely used as NMR solvents. Since the deuterium isotope of hydrogen has a different magnetic moment and spin, it is invisible in a spectrometer tuned to protons.

Spin–Spin Coupling

Consider the structure of ethanol:

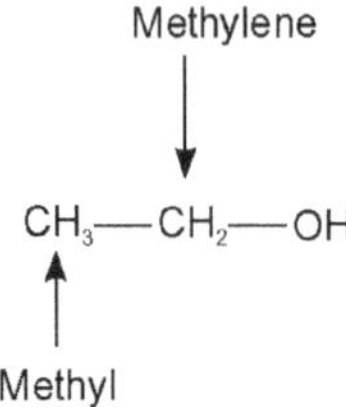

The ^{1}H NMR spectrum of ethanol (Figure 4.5) shows that the methyl peak has been split into three peaks (a **triplet**), and the methylene peak has been split into four peaks (a **quartet**). This occurs because there is a small interaction (**coupling**) between the two groups of protons. The spacing between the peaks of the methyl triplet is equal to the spacing between the peaks of the methylene quartet. This spacing is measured in hertz and is called the coupling constant J.

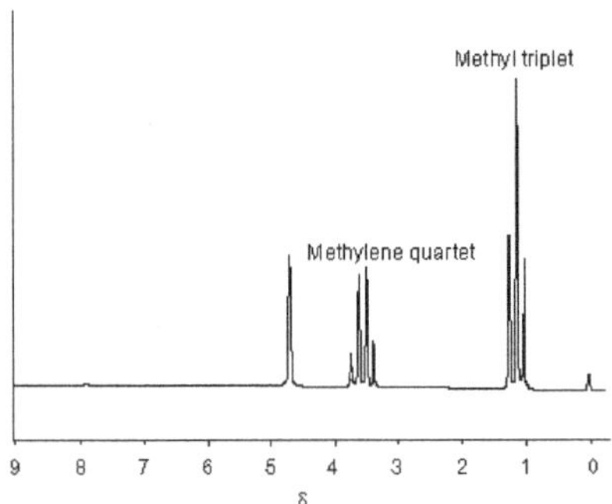

Figure 4.5   NMR spectrum of ethanol

To see why the methyl peak is split into a triplet, let us look at the **methylene** protons. There are two of them, and each can have one of the two possible orientations (aligned with or opposed against the applied field). This gives a total of four possible states (Figure 4.6).

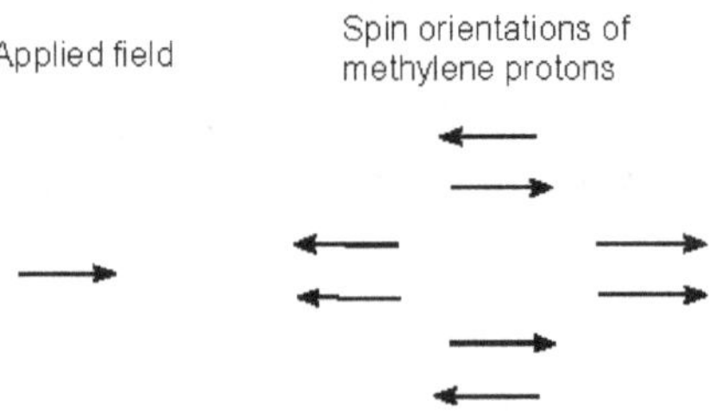

Figure 4.6   Four possible orientations of methylene protons

In the first possible combination, spins are paired and opposed to the field. This has the effect of reducing the field experienced by the **methyl** protons; therefore, a slightly higher field is needed to bring them to resonance, resulting in an upfield shift. Neither combination of spins opposed to each other has an effect on the methyl peak. The spins paired in the direction of the field produce a downfield shift. Hence, the methyl peak is split into three with the ratio of areas 1: 2 :1.

Similarly, the effect of methyl protons on methylene protons is such that there are eight possible spin combinations for the three methyl protons (Figure 4.7).

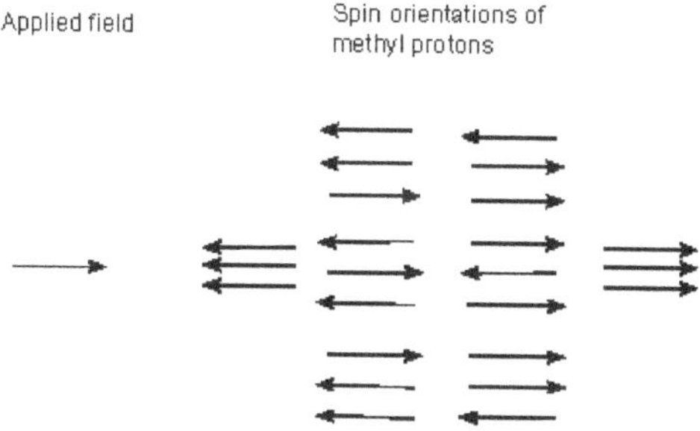

Figure 4.7   Eight possible spin combinations of methyl protons

Out of these eight groups, there are two groups of three magnetically equivalent combinations. The methylene peak is split into a quartet. The areas of the peaks in the quartet have the ratio 1: 3: 3: 1.

By extending these arguments to larger numbers of spins, it is possible to show that N equivalent spins split the resonance of a coupled group into N+1 lines, with intensities corresponding to the coefficients of a **binomial expansion** or **Pascal's triangle** (Figure 4.8).

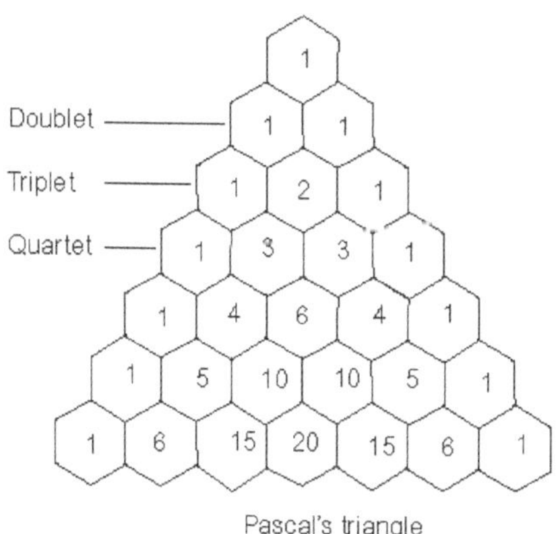

Figure 4.8   Pascal's triangle

In a first-order spectrum (where the chemical shift between interacting groups is much larger than their coupling constant), interpretation of splitting patterns is quite straightforward.

- The multiplicity of a multiplet is given by the number of equivalent **protons** in **neighbouring** atoms plus one, i.e., the $n + 1$ rule.

- Equivalent nuclei do not interact with each other. The three methyl protons in ethanol cause splitting of the neighbouring methylene protons; they do not cause splitting among themselves.

- The coupling constant is not dependent on the applied field. Multiplets can be easily distinguished from closely spaced chemical shift peaks.

Inductive Effect

If the electron density about a proton nucleus is relatively high, the induced field due to electron motion will be stronger than that when the electron density is relatively low. The shielding effect in such high electron density cases will, therefore, be larger, and a higher external field will be needed for the RF energy to excite the nuclear spin. Since silicon is less electronegative than carbon, the electron density about the methyl hydrogens in tetramethylsilane is expected to be greater than the electron density about the methyl hydrogens in neopentane (2,2-dimethylpropane), and the characteristic resonance signal from the silane derivative indeed lies at a higher magnetic field. Such nuclei are said to be **shielded**. Elements that are more electronegative than carbon should exert an opposite effect (reduce the electron density); methyl groups bonded to such elements display lower field signals (they are **deshielded**). Thus chemical shift values increase from 0.9 to 4.1 as the compound is changed from $(CH_3)_4C$ to $(CH_3)_4F$. Similar increases can be found when $(CH_3)_4Si$ is changed to CH_3Cl from 0.0 to 3.0. The δ values of the following compounds are as follows:

Compound	$(CH_3)_4C$	$(CH_3)_3N$	$(CH_3)_2O$	CH_3F
δ (ppm)	0.9	2.1	3.2	4.1

Compound	$(CH_3)_4Si$	$(CH_3)_3P$	$(CH_3)_2S$	CH_3Cl
δ (ppm)	0.0	0.9	2.1	3.0

The deshielding effect of electron-withdrawing groups is roughly proportional to their electronegativity. Furthermore, if more than one such group is present, the deshielding is additive and proton resonance is shifted even further downfield (Table 4.3).

Table 4.3 Proton chemical shifts (ppm) of substituted methane

Cpd./Sub					
CH_3X	3.0	2.7	2.1	3.1	2.1
CH_2X_2	5.3	5.0	3.9	4.4	3.7
CHX_3	7.3	6.8	4.9	5.0	4.3

The OH proton signal is seen at 2.37 δ in 2-methyl-3-butyne-2-ol, and at 3.87 δ in 4-hydroxy-4-methyl-2-pentanone, illustrating the wide range over which this chemical shift may be found. The intramolecular hydrogen bonding in 4-hydroxy-4 methyl-2-pentanone is partly responsible for the shift of δ to low field. We can take advantage of the rapid OH exchange with the deuterium of heavy water to assign hydroxyl proton resonance signals. As shown in the following equation, this removes the hydroxyl proton from the sample and its resonance signal in the NMR spectrum disappears. Experimentally, one simply adds a drop of heavy water to a chloroform-d solution of the compound and runs the spectrum again. The result of this exchange is shown in Figure 4.9.

Pi-electrons are more polarizable than sigma-bond electrons, as addition reactions of electrophilic reagents to alkenes testify. Therefore, we should not be surprised to find that field-induced pi-electron movement produces strong secondary fields that perturb nearby nuclei. The pi-electrons associated with a benzene ring provide a striking example of this phenomenon, as shown in Figure 4.10. The electron clouds above and below the plane of the ring circulate in reaction to the external field so as to generate an opposing field at the centre of the ring and a supporting field at the edge of the ring.

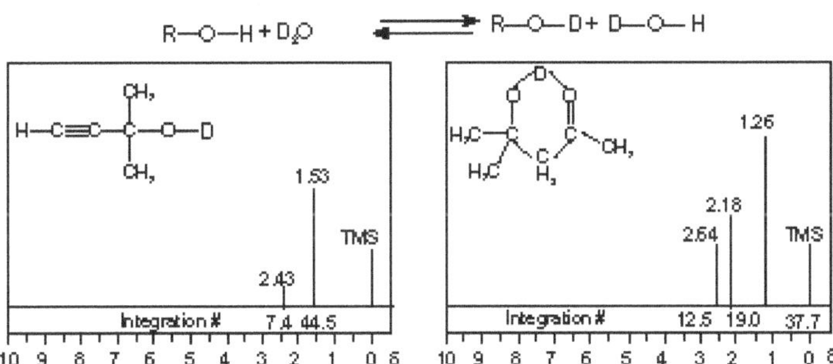

Figure 4.9 Hydrogen bonding shifts the resonance signal of a proton to lower field (higher frequency)

This kind of spatial variation is called **anisotropy**, and it is common to non-spherical distributions of electrons. Regions in which the induced field supports or adds to the external field are said to be **deshielded**, because a slightly weaker external field will bring about resonance for nuclei in such areas. However, regions in which the induced field opposes the external field are termed **shielded** because an increase in the applied field is needed for resonance. Shielded regions are designated by a **plus sign**, and deshielded regions by a **negative sign**. Note that anisotropy about the triple bond nicely accounts for the relatively high field chemical shift of ethynyl hydrogens. The shielding and deshielding regions about the carbonyl group have been described in two ways, which alternate in the display.

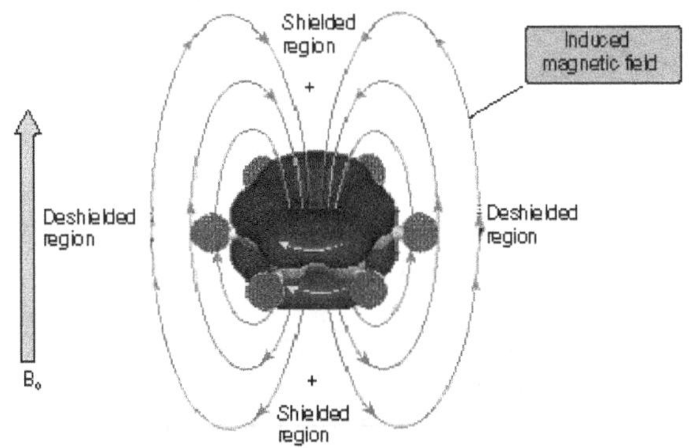

Figure 4.10   Shielding and deshielding in benzene

Continuous wave NMR spectrometer

A typical continuous wave (CW) spectrometer is shown in Figure 4.11. A solution of the sample in a uniform 5-mm glass tube is oriented between the poles of a powerful magnet, and is spun to average any magnetic field variations as well as tube imperfections. Radio frequency radiation of appropriate energy is broadcast into the sample from an antenna coil. A receiver coil surrounds the sample tube, and emission of absorbed RF energy is monitored by dedicated electronic devices and a computer.

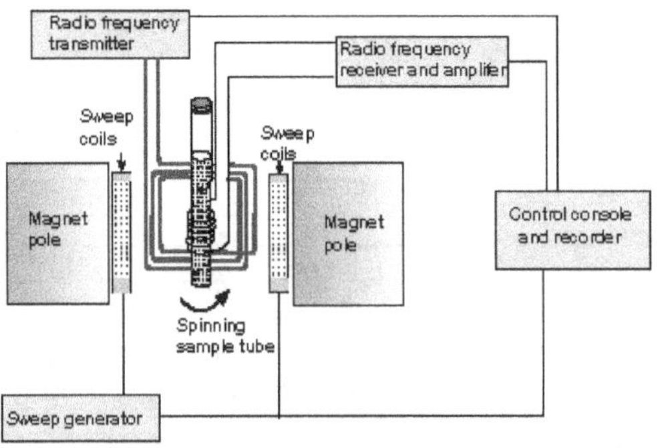

Figure 4.11   Continuous wave spectrometer

The Magnet

The NMR magnet is one of the most expensive components of the nuclear magnetic resonance spectrometer. Most magnets are of the superconducting type. A superconducting magnet is an electromagnet made of superconducting wire. Superconducting wire has a resistance approximately equal to zero when it is cooled to a temperature close to absolute zero ($-273.15°C$ or 0 K) by immersing it in liquid helium. Once current is caused to flow through the coil, it will continue to flow as long as the coil is kept at liquid helium temperatures.

Shim Coils

The purpose of shim coils in a spectrometer is to correct minor spatial variations in the homogeneity in the B_o magnetic field. These variations could be caused by the magnet design, materials in the probe, variations in the thickness of the sample tube, sample permeability, and ferromagnetic materials around the magnet. A shim coil is designed to create a small magnetic field that will oppose and cancel out any variation in homogeneity in the B_o magnetic field.

Sample Probe

The sample probe is a part in the spectrometer that accepts the sample, sends RF energy into the sample, and detects the signal emanating from the sample. It contains the RF coil, sample spinner, temperature controlling circuitry and gradient coils.

Sample Spinner

The purpose of the sample spinner is to rotate the NMR sample tube about its axis. In doing so, each spin in the sample located at a given position along the z-axis and radius from the z-axis will experience average magnetic field in the circle defined by Z and its radius. The net effect is a narrower spectral line width.

Radio Frequency Oscillator

The RF coils create a field that rotates the net magnetization in a pulse sequence. They also detect transverse magnetization as it precesses in the XY plane.

Each of these RF coils must resonate, i.e., they must efficiently store energy at the Larmor frequency of the nucleus being examined with NMR spectrometer. All NMR coils are composed of an inductor, or inductive elements, and a set of capacitors elements.

Recorder and Integrator

In routine analysis, the signal is sent directly to the recorder or oscilloscope. The oscilloscope is useful for quick scanning of the spectrum and allows for preliminary adjustments to be made, which include setting optimum sensitivity or gain and electronic filtering of the signal. The recorder records the spectrum as a plot of resonance signal on the y-axis versus the strength of magnetic field on the x-axis. Most spectrometers are equipped with integrators to measure the area under the observed signal. Such an area is proportional to the number of resonating nuclei (Figure 4.12).

An NMR spectrum is acquired by varying or sweeping the magnetic field over a small range while observing the RF signal from the sample. An equally effective technique is to vary the frequency of the RF radiation while holding the external field constant.

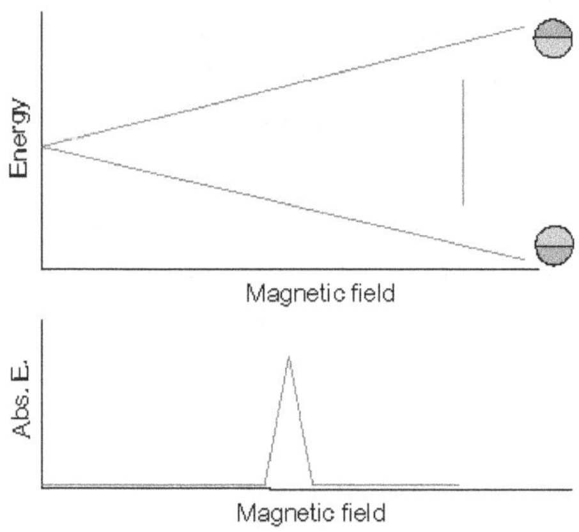

Figure 4.12   Relationship between magnetic field strength and transition energy

Since protons have the same magnetic moment, we might expect all hydrogen atoms to give resonance signals at the same field/frequency values. Fortunately, for chemical applications, this is not true. This is due to the effect of electron(s) surrounding the proton in covalent compounds and ions. Since electrons are charged particles, they move in response to the external magnetic field (B_o) so as to generate a secondary field that opposes the much stronger applied field. This secondary field **shields** the nucleus from applied field, and so B_o must be increased in order to achieve resonance (absorption of RF energy). As shown in Figure 4.12, B_o must be increased to compensate for the induced shielding field. In the figure, the more shielded nuclei (HCl, HBr and HI) appear at higher field side.

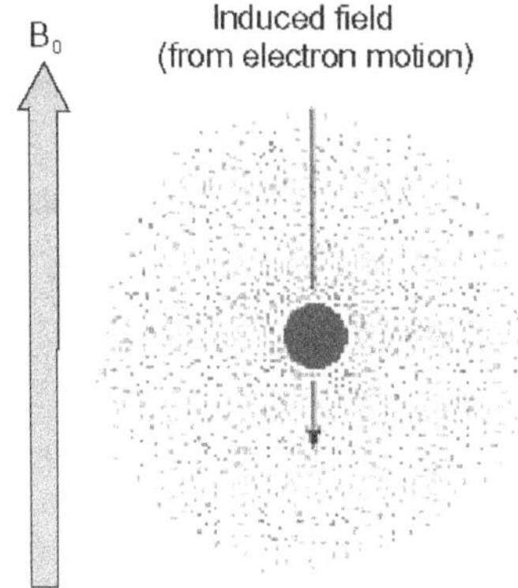

Figure 4.13   Shielding by secondary field

The magnetic field range displayed in Figure 4.13 is very small as compared with the actual field strength (only about 0.0042%). It is customary to refer to such small increments in units of **parts per million** (ppm). The difference between 2.3487 T and 2.3488 T is, therefore, about 42 ppm. Instead of designating a range of NMR signals in terms of magnetic field differences, it is more common to use a frequency scale, even though the spectrometer may operate by sweeping the magnetic field. Using this terminology, we would find that at 2.34 T, the proton signals extend over a 4,200 Hz range (i.e., for a 100 MHz RF frequency, 42 ppm means 4,200 Hz). Most organic compounds exhibit proton resonances that fall within a 12-ppm range and it is therefore necessary to use very sensitive and precise spectrometers to resolve structurally distinct sets of hydrogen atoms within this narrow range. In this respect, it might be noted that the detection of ppm difference is equivalent to detecting a 1-millimetre difference in a distance of 1 kilometre.

Pulsed Fourier Transform NMR spectroscopy

A far better resolution and sensitivity in NMR may be achieved by the introduction of pulsed Fourier transform techniques (FT–NMR). In FT–NMR, the resonances are not measured one after another, but all nuclei are excited at the same time by a radio frequency pulse. Normally, a radio emitter works at a fixed frequency. However, if the radiation is emitted as a very short pulse (some µs in NMR), the pulse frequency becomes 'uncertain'. A short radio frequency pulse contains many frequencies in a broad band and thus excites the resonances of all spins in a sample at the same time.

The excited spins emit the absorbed radiation after the pulse. The emitted signal is a superposition of all excited frequencies. Its evolution in time is recorded. The intensities of several frequencies, which give the observed signal in their superposition, are calculated by a mathematical operation (the Fourier transformation), which translates the time data into the frequency domain. The resulting NMR spectrum looks like an ordinary CW spectrum but its resolution is several orders of magnitudes better.

To obtain good spectra free from too much electronic noise, it is common to add together between 8–800 FIDs, a process which takes between ~30 seconds to 24 hours. During this period the field frequency may drift slightly, resulting in poor averaging. To compensate this, most FT spectrometers have special "lock circuitry" based on detecting a deuterium signal. For this to work, the solvent used must contain deuterium! Normally $CDCl_3$ is used, but deuterated acetone or DMSO are also common. "Locking" the sample is normally the first operation actually performed on the spectrometer. The "lock signal" is then used to "shim" the spectrometer, i.e., adjust the homogeneity of B_o.

At the end of the short radio frequency pulse t_p, the precession of all the nuclei in the magnetic field will be effectively in phase. If the measurement of the resultant induced signal starts immediately, all the initial measured sine wave responses would also be in phase. However, no NMR spectrometer can achieve this. For electronic reasons, a short delay between the end of t_p and start of measurement is required. The short delay means measurement begins when the sine waves are already out of phase. This so-called **first order phase error** can be calculated from a suitable combination of the real and imaginary components of $F(\omega)$ Another phasing error due to technical imperfections in the spectrometer is often referred to as the **zero order phase** correction. These are both applied to $F(\omega)$ after the FT is complete.

With the measurement of a single FID resulting from one pulse, followed by Fourier transformation to give $F(\omega)$ one can achieve a spectrum approximately equivalent in its noise level to a CW spectrum. By adding n FID measurements together in the computer, one can reduce the noise by $\hat{}n$, i.e., 64 FIDs added together (and taking 3.41*64 = 3.6 minutes) will reduce the noise by almost a factor of 10. It also turns out that one can multiply the whole of f(t) by a new function such as a decaying exponential prior to the FT operation. Surprisingly, this does NOT introduce any new frequencies, but can dramatically reduce the noise. Hence the resulting peaks broader. This is because a faster f(t) decay corresponds to a shorter relaxation time T, a larger $1/T$, and hence wider peaks.

Other more complex functions can actually decrease apparent peak width. The effects of such functions, result in the following output as shown Figure 4.14.

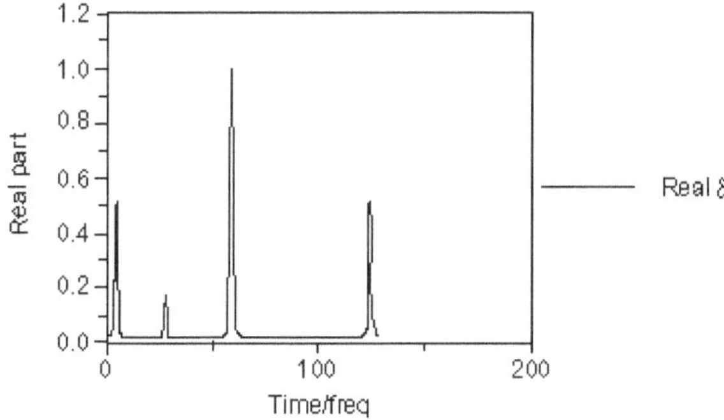

Figure 4.14   FT-NMR spectrum

Carbon-13 NMR spectroscopy

Many of the molecules studied by NMR contain carbon. Unfortunately, the carbon-12 nucleus does not have a nuclear spin, but the carbon-13 (C-13) nucleus has a nuclear spin due to the presence of an unpaired neutron. Carbon-13 nuclei make up approximately one per cent of the carbon nuclei on earth. Therefore, carbon-13 NMR spectroscopy will be less sensitive (have a poorer SNR) than hydrogen NMR spectroscopy. With appropriate concentration, field strength and pulse sequences, however, carbon-13 NMR spectroscopy can be used to supplement the previously described hydrogen NMR information. Advances in superconducting magnet design and RF sample coil efficiency have helped make carbon-13 spectroscopy routine in most NMR spectrometers.

Applications of ^{13}C NMR

To illustrate some of the concepts described earlier, we will first concentrate on carbon. The most abundant isotope ^{12}C has no overall nuclear spin, having an equal number of protons and neutrons. The ^{13}C isotope, however, does have a spin 1/2 but is only 1% abundant. Carbon NMR spectra are characterized by the following:

- a chemical shift range of about 220 ppm normally expressed relative to the ^{13}C resonance of TMS;
- a natural line width of approximately 1 Hz related to the values of the relaxation times T_1 and T_2;
- a Larmor frequency in the range of 20–100 MHz for typical spectrometers;
- about 5–20 mg of sample dissolved in 0.4–2 ml of solvent (normally $CDCl_3$) is required, and a good spectrum would be obtained in 64–6400 scans.

Let us begin with a ^{13}C spectrum of diethyl phthalate obtained by the FT technique (Figure 4.15).

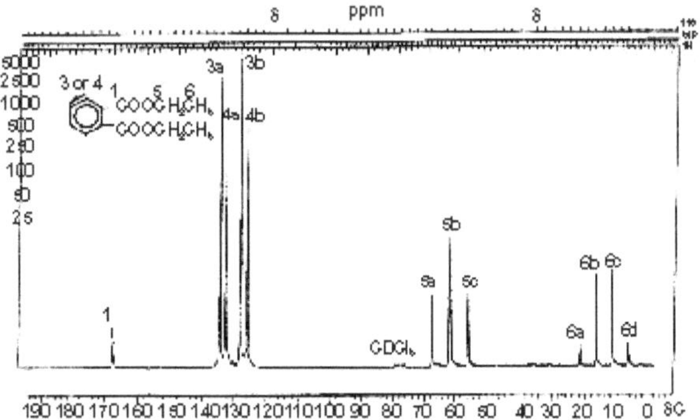

Figure 4.15 ^{13}C NMR spectrum of diethyl phthalate with the protons completely coupled. CDCl$_3$ solvent, 25.2 MHz

First of all, note the wide chemical shift range of the signals. Note also that the signal that is attributed to the methyl group is approximately a 1:3:3:1 quartet; for methylene approximately a 1:2:1 triplet; and for aromatic CHs approximately 1:1 doublets. The intensity ratios suggest that this coupling and multiplicities due specifically to coupling of the spin 1/2 ^{13}C with the spin 1/2 protons and nothing else (i.e., the 2nI+1 rule, I being the spin number). Before we move on to discuss how this coupling may be useful, let us remind ourselves of how the coupling arises. Recall the energy level diagram of one spin 1/2 nucleus in a magnetic field (Figure 4.16).

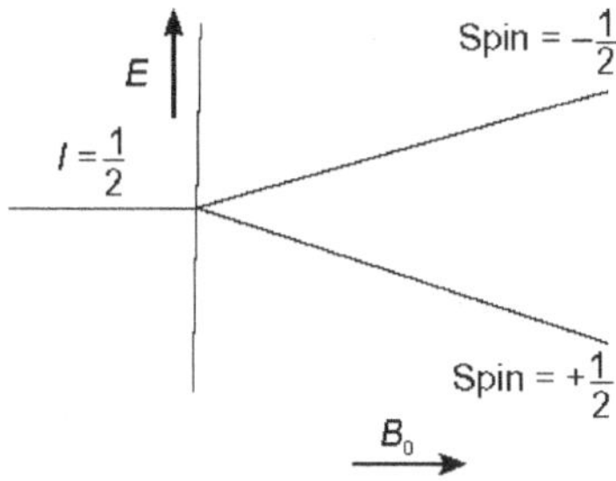

Figure 4.16   Energy level diagram for a nucleus with spin 1/2

The precessing magnetization vector either reinforces or opposes B_o so that locally, at least, other nuclei will perceive two slightly different values of B_o. Since the populations of each energy level are practically identical another close-by nucleus will resonate with equal probability at two slightly different Larmor frequencies. The difference between these two frequencies is what we know as the coupling constant, J. Its value depends on how the perturbation in B_o is transmitted between the two nuclei, and this is normally achieved via the intervening electrons (hence the term through bond coupling). When two identical nuclei are involved, three slightly different and equally spaced values of B_o with 1:2:1 triplet coupling pattern are produced. In spin terms, we say that four configurations are possible: +1/2,+1/2; +1/2, −1/2; −1/2, +1/2; −1/2, −1/2. As the middle two are of equal energy, this would manifest as a double height peak, i.e., a 1:2:1 triplet. Carbon appears not to couple with other carbons, but only with protons in the same molecule. This is because the probability that two $^{13}C–^{13}C$ nuclei will be close enough to couple is 100 times less than the probability of finding $^{13}C–^{12}C$ as adjacent nuclei.

The decoupled spectrum for diethyl phthalate is shown in Figure 4.17. There are seven singlets; the spectrum is certainly less cluttered because the coupling has gone! It has been removed by a technique called **broad band** ^1H decoupling. During measurement of the ^{13}C FID, the entire proton resonance region of the compound (i.e., from about −2 to +10 ppm) was irradiated with "white noise" radio frequency. This has the effect of increasing the rate of transition between the proton low and high energy spin states such that only one average magnetic field is experienced by any individual carbon nucleus, and the manifestation of coupling vanishes. As far as the carbon is concerned, all the protons cease to exist. There is an effect called the nuclear Overhauser effect (NOE), which does very odd things to the intensity of the carbon line, normally increasing it by a factor of two or more, and hence is one of the factors making integration of ^{13}C spectra unreliable. Normally, for ^{13}C, both decoupled and coupled spectra are recorded, but the latter with a small modification called **off-resonance proto decoupling**.

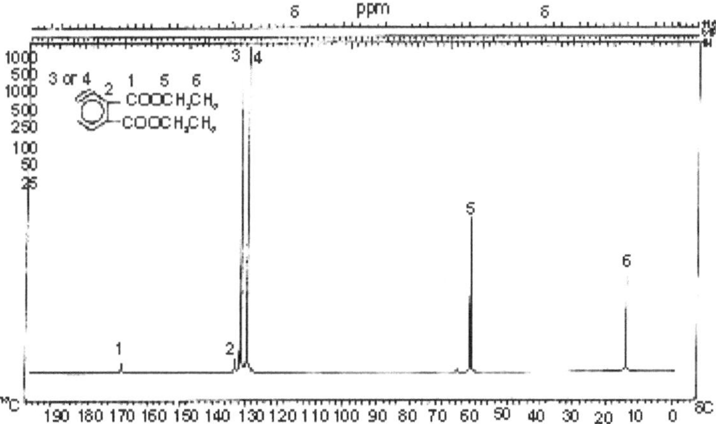

Figure 4.17 ^{13}C NMR spectrum of diethyl phthalate with the protons completely decoupled by broadband noise, CDCl₃ solvet, 25.2 MHz.

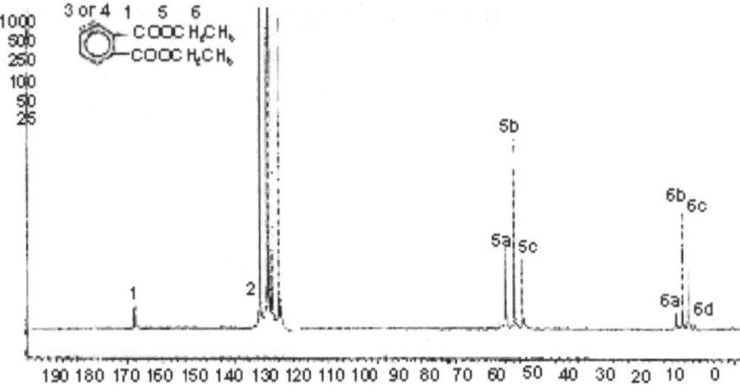

Figure 4.18   The ^{13}C NMR spectrum of diethylphthalate with off resonance decoupled protons

Here, the proton decoupler is actually switched on during ^{13}C measurement but its frequency is centred at about −10 ppm. It has the effect of removing any coupling between ^{13}C and ^{1}H that occurs through more than one bond (H—C—C and longer range couplings of <10 Hz) but leaving coupling due to directly bonded atoms (i.e., H—C of >125 Hz) still visible. This reduces the complexity of the spectrum and offers a significant gain in the intensity of signals (approximately fourfold reduction in measurement time) from the NOE enhancement referred to earlier (Figure 4.18).

Together, these two spectra give the following information:

- The proton decoupled spectrum tells the number of unique types of carbon atoms in the molecule (i.e., a mono-substituted phenyl group has **four** unique carbon atoms).

- The off-resonance spectrum tells how many hydrogen atoms are attached to each unique carbon (quartet = 3, triplet = 2, doublet = 1, singlet = none).

- From the chemical shift of each carbon, much information about the environment of the carbon can be gleaned.

The typical chemical shift ranges of carbon nuclei areas follows:

C (alkane)	~0–30 ppm
C (alkene)	~110–150 ppm
C—N	~50
C—O	~60
C—F	~70 ppm
Aromatic	~110–160 ppm
Ester, amide, acid	~ 160–170 ppm
Ketone, aldehyde	~ 200–220 ppm

Two-dimensional NMR

A two-dimensional NMR experiment involves a series of one-dimensional experiments. Each experiment consists of a sequence of radio frequency pulses with delay periods in between them. It is the timing, frequencies, and intensities of these pulses that distinguish different NMR experiments from one another. During some of the delays, the nuclear spins are allowed to freely precess (rotate) for a determined length of time known as **evolution time**. The frequencies of nuclei are detected after the final pulse. By incrementing the evolution time in successive experiments, a two-dimensional data set is generated from a series of one-dimensional experiments Figure 4.19).

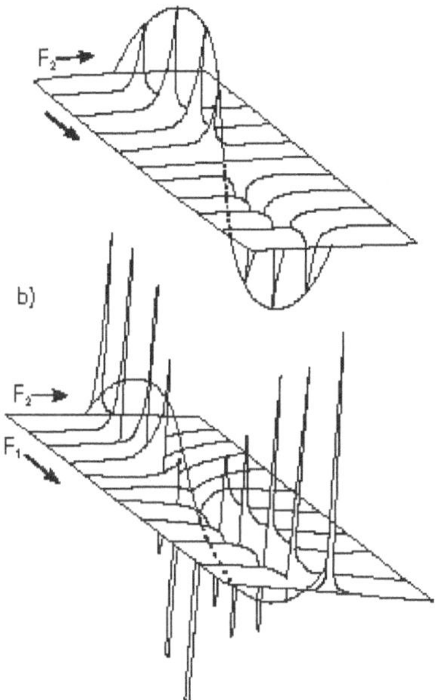

Figure 4.19   Generation of two-dimensional data set

An example of a two-dimensional NMR experiment is the homonuclear correlation spectroscopy (COSY) sequence, which consists of a pulse (p_1) followed by an evolution time (t_1) followed by a second pulse (p_2) followed by a measurement time (t_2). A computer is used to compile the spectra as a function of evolution time (t_1). Finally, the Fourier transform is used to convert the time-dependent signals into a two-dimensional spectrum.

The two-dimensional spectrum that results from the COSY experiment shows the frequencies for a single isotope (usually hydrogen, ^1H) along both axes. Several techniques have also been devised for generating heteronuclear correlation spectra, in which the two axes correspond to different isotopes, such as ^{13}C and ^1H. The intensities of the peaks in the spectrum can be represented using a third dimension.

More commonly, intensity is indicated using contours or different colours. The spectrum is interpreted starting from the diagonal, which consists of a series of peaks. The peaks that appear off the diagonal are called cross-peaks. Cross-peaks are symmetrical (both above and below the diagonal) and indicate which hydrogen atoms are spin–spin coupled to each other. One can determine which atoms are connected to one another by only a few chemical bonds by matching the centre of a cross-peak with the centre of each of two corresponding diagonal peaks. The peaks on the diagonal when matched with cross-peaks are coupled to each other.

For example, a $CH_3CH_2COCH_3$ molecule (2-butanone) would show three peaks on the diagonal due to the three distinct hydrogen groups. By drawing a line straight down from a cross-peak to the point on the diagonal directly above or below it, and then drawing a line from the cross-peak directly across to another peak on the diagonal, one can determine which peaks are coupled. This is done in such a way that the lines from the cross-peak form a 90° angle between the two peaks on the diagonal. The matching peaks, as determined by using the cross-peaks, indicate which hydrogens are coupled, giving a clearer understanding of the structure of the molecule under examination.

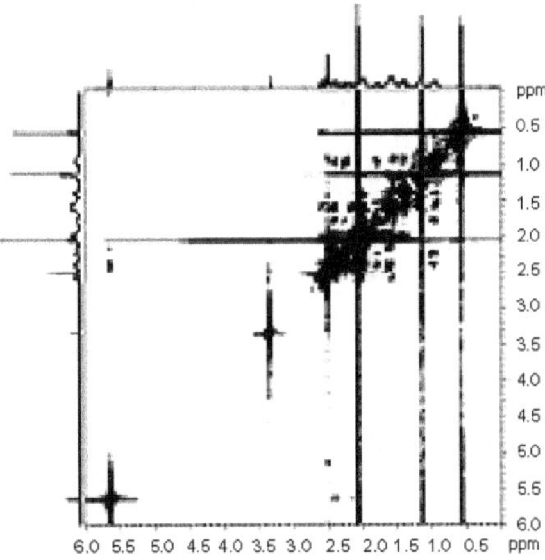

Figure 4.20   2D proton COSY experiment on progesterone

An example of a COSY NMR spectrum of progesterone in DMSO-d6 is shown in Figure 4.20. The spectrum that appears along both the x- and y-axes is a regular one-dimensional ^{1}H NMR spectrum. The COSY is read along the diagonal where the bulk of peaks appear. Cross-peaks appear symmetrically above and below the diagonal.

COSY NMR

COSY-90 is the most common COSY experiment. In COSY-90, the sample is irradiated with a radio frequency pulse, p_1, which tilts the nuclear spin by 90°. After p_1, the sample is allowed to freely precess during an evolution period (t_1). A second 90° pulse, p_2, is then applied after which the experimental data are acquired. This is done repeatedly using a series of different evolution periods (t_1). At the conclusion of data acquisition, the data is Fourier transformed in each dimension to generate the two-dimensional spectrum (Figure 4.21). It is only because the evolution period is varied that cross-peaks appear in the spectrum.

Figure 4.21 COSY-90 spectrum

In COSY NMR, the resonance signal from the sample is read in time period t_2 following two experimental magnetic pulses p_1 and p_2 separated by a variable time period t_1.

Cross-peaks result from a phenomenon called **magnetization transfer**. Depending on the experiment, this transfer can be achieved through space or bonds, or even through chemical or physical means. In COSY, magnetization, transfer occurs through the bonds.

Another member of the COSY family is **COSY-45**. In COSY-45, a 45° pulse is used instead of a 90° pulse for the first pulse, p_1. The advantage of a COSY-45 is that the diagonal peaks are less pronounced, making it simpler to match cross-peaks near the diagonal in a large molecule. Additionally, the relative signs of coupling constants can be elucidated from a COSY-45 spectrum. This is not possible using COSY-90. Overall, COSY-45 offers a cleaner spectrum while COSY-90 is more sensitive. Related COSY techniques include double quantum filtered COSY and multiple quantum filtered COSY.

COSY NMR has many useful applications. Organic chemists often use COSY to elucidate structural data on molecules that are not satisfactorily represented in a one-dimensional NMR spectrum. Using cross-peaks, along with the diagonal spectrum, one can discover much about the structure of an unknown molecule.

Double Resonance and NOE

Double Resonance Experiments

In the case of usual NMR experiment with frequency sweep, a variable field B_1 is employed for the observation of the absorption spectrum. Double resonance is the term applied to experiments in which a field B_2, in addition to B_1, is applied to the sample. The result of such an experiment can vary widely, depending upon the frequency and amplitude of B_2.

Spin Decoupling

The double resonance technique is experimentally the simplest in the frequency sweep mode. In this case, B_2 field is produced at the desired position in the spectrum by side-band modulation from the central band of the oscillator, and the entire absorption spectrum can be observed using a second side band of variable frequency as B_1 field. Thus, in a single experiment, all interactions given by the nucleus be eliminated and assigned. The extension of double resonance to triple resonance by the use of a third field B_3 can be accomplished without difficulty. The double resonance technique can also be applied in the field-sweep mode, but for complete analyses of complicated spectra, several experiments are necessary. The second field is produced through a side band at constant separation Δv from the field B_1. In practice, the decoupling effect is observed only when Δv is equal to the difference between the resonance frequencies of the decoupled and observed nucleus. Double resonance technique can also be used to determine resonance frequencies even when their signals cannot be identified because of superposition with the absorption lines of other nuclei.

Spin Tickling and Selective Double Resonance

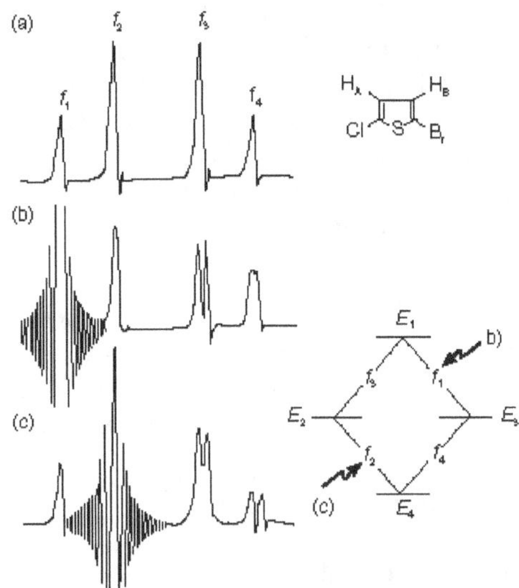

Figure 4.22 Spin tickling experiment in the AB system of 2-bromo-5-chlorothiophene

In common double resonance technique, B_2 fields of relative high amplitude are necessary in order to achieve the desired effect of decoupling. Other effects are observed if weak B_2 fields are used. In fact, one irradiates the corresponding transition in the energy level diagram if B_2 has the frequency of a particular resonance signal. This leads to a splitting of every line in the spectrum that has an energy level in common with the perturbed transition. In the case of AB one may expect that an irradiation of line f_1 would result in splitting of lines f_3 and f_4 while line f_2 would remain unchanged (Figure 4.22).

A plausible explanation for this phenomenon known as spin tickling experiment states that, as a result of perturbation, the eigen values, E_1 and E_3, of the spin system mix with one another and, therefore, two transitions become possible. The new transition practically corresponds to the previously forbidden double quantum transition $E_4 E_1$. It is noteworthy that the way in which the transitions are connected in the energy level diagram manifests itself in the experiment. We differentiate the progressively connected transitions in which the three eigen values continuously change their energies and regressively connected transitions in which the central eigen value is larger or smaller than the energy of the initial and final states. Initial and final states of a progressively linked line pair differ in their total $\Delta m_\gamma = -2$. spin by two units: On the other hand, for regress $\Delta m_\gamma = 0$. vely linked pairs,

Now, we experimentally observed that regressively linked transitions are sharply split (or resolved), whereas in progressively linked transitions the splitting is less well defined (Figure 4.23). The spin tickling process thus provides an elegant possibility for probing the energy level diagram with reference to an experimental spectrum. It is, therefore, an important aid in the analysis of spectra, and is particularly useful for the determination of the relative signs of spin–spin coupling constants, in systems with more than two nuclei.

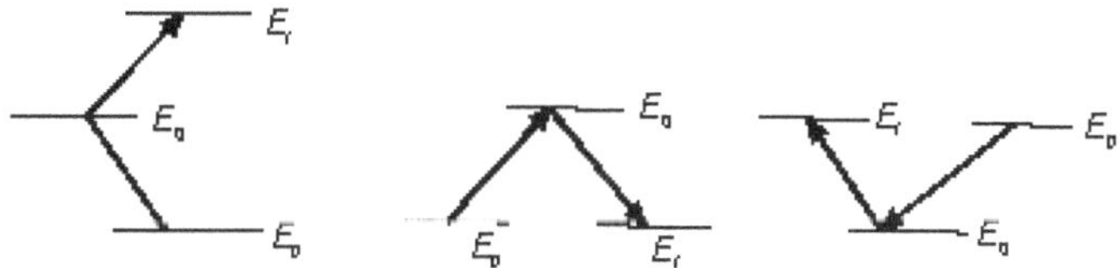

Figure 4.23   Regressively linked transitions

The Overhauser Effect

Three different phenomena are usually considered in connection with the notion of the Overhauser effect. Each involves a variation of the intensity of resonance signals observed in double resonance experiments. However, several different mechanisms are responsible for the effect.

In the case of true Overhauser effect for a system consisting of a nuclear spin I and an electron spin S, an increase in the intensity of nuclear resonance signal is observed if one simultaneously saturates the electron resonance with an RF field of frequency v_s. This experiment may be performed on a paramagnetic solution of sodium in liquid ammonia in which proton resonance is observed under conditions that saturate the electron resonance.

The variation of the intensities of nuclear resonance lines that result from this experiment can be rationalized by referring to the Solomon diagram Figure 4.24. There the eigen states of a two-spin system (IS) in a magnetic field are represented. Altogether, there exist four states of different energy, and for their arrangement different signs of the nuclear and electron spins were considered. Transitions for the nucleus or electron can be stimulated by an RF field of frequency v_1 or v_2, respectively. Let us now consider the probability, W, for the particular relaxation transitions that are responsible for the maintenance of Boltzmann distribution. Thus, the quantities W_1 and W'_1 correspond to the probability for longitudinal relaxation of nuclear and electron spins, respectively. In addition, there are also transition probabilities W_2 and W_0 for cases in which the nuclear and electron spins flip simultaneously.

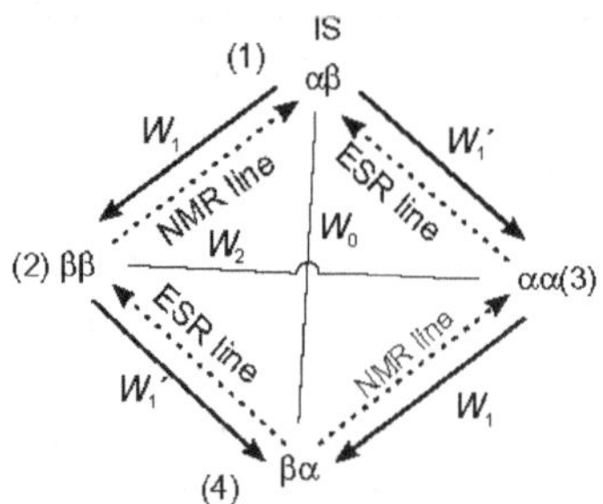

Figure 4.24   Solomon diagram for a two-spin system (*IS*) composed of a nuclear and an electron spin

W_2 and W_0 are of significance only when there is a spin–spin interaction between the spins I and S. Now, if the electron resonance, i.e., transitions (3)→(1) and (4)→(2), is saturated by an oscillating field B_1 with frequency v_s, then the Boltzmann distribution between the states (3) and (1) as well as between (4) and (2) will be disturbed, i.e., the populations of (1) and (2) will become too high while those of (3) and (4) will become too low. This perturbation can be counteracted by an increased number of relaxation transitions, i.e., through an increase in W_0, since, in this fashion, the state (1) is depopulated and the population of state (4) is increased. However, for nuclear resonance, i.e, for transitions (4)→(3) and (2)→(1), this leads to an intensification of the absorption signal because the net effect of the process is the overpopulation of state (2) and the depopulation of state (3) since the spins are carried along the route (3)→(1)→(4)→(2). The result is polarization of the nuclear spin distribution, and thus the experiment is known as dynamic nuclear polarization.

It is important for the preceding case that the relation $W_0 \gg W_2$ is satisfied. An electron spin, thus, can flip only when the nucleus simultaneously changes its spin orientation in the opposite direction. In this case, the relaxation is produced predominantly through a time-dependent scalar spin–spin coupling. In the previously mentioned solution of sodium in liquid ammonia, the unpaired electrons are solvated by ammonia molecules. A fast exchange of molecules between the solvation shells of different paramagnetic centres has the result that the proton electron coupling vanishes. It maintains its effectiveness, however, as a relaxation mechanism.

Let us revisit the experiment we have just described but with reference to the magnitude of energies that are exchanged with the lattice in the relaxation process. For every quantum mechanical system that is characterized by two energy levels E_p and E_q, equilibrium is established so that the number of transitions $E_p \to E_q$ is equal to the number of transitions $E_q \to E_p$. Thus, it follows that for the eigen states (1) and (4) of the spin system (IS),

$$N_\alpha n_\beta W_{\alpha\beta \to \beta\alpha} = N_\beta n_\alpha W_{\beta\alpha \to \alpha\beta}$$

in which N_α, N_β and n_α, n_β signify the populations for the nuclei and electros, respectively, and $W_{\alpha\beta \to \beta\alpha}$ and $W_{\beta\alpha \to \alpha\beta}$ denote the transition probabilities. According to the Botzmann law, we have

$$\frac{N_\alpha n_\beta}{N_\beta n_\alpha} = \frac{W_{\beta\alpha \to \alpha\beta}}{W_{\alpha\beta \to \beta\alpha}} = \exp\left(\frac{-\Delta E}{kT}\right) = \exp[h(v_a + v_1)/kT] \tag{4.1}$$

Now, if the electron resonance is saturatd, then $n_\alpha = n_\beta$ and

$$\frac{N_\alpha}{N_\beta} = \exp(h(v_s + v_I)/kT) \tag{4.2}$$

Because $hv_s \gg hv_I$, the nuclear spin distribution, which normally beys the expression

$$\frac{N_\alpha}{N_\beta} = \exp(hv_I/kT)$$

is now determined by very much larger energy difference hv_s.

If we extend this discussion to a spin system that consists of two nuclear spins, we arrive at the so-called nuclear Overhauser effect (NOE). For this case, the Solomon diagram must be modified, since both spins have the same sign and the sequence of the states is changed (Figure 4.25).

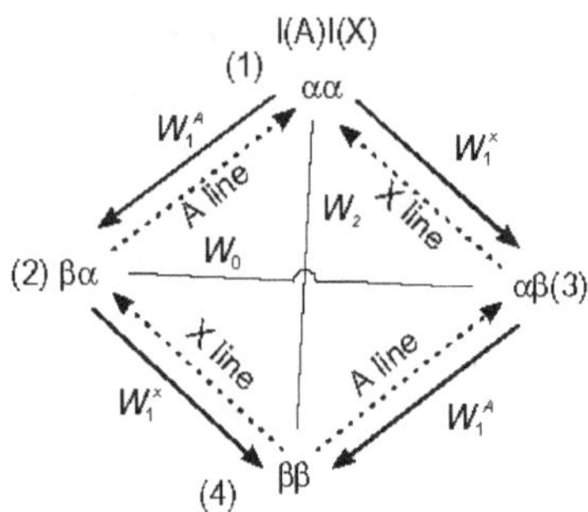

Figure 4.25   Solomon diagram for a two-spin system (II) composed of two nuclear spins.

Now, if the resonance of one nucleus, for example the X nucleus in Figure 4.26 is irradiated, an increase in the intensity of resonance of A nucleus occurs if $W_0 <<$ W_2. This condition is met in all cases in which a time-dependent dipolar coupling is responsible for relaxation.

The quantitative treatment of the phenomenon leads to the so called Solomon equation that yields an expression for the increase in the z-magnetization of nucleus A, M_z^A, relative to the equilibrium magnetization, M_0^A, caused by the seondary field at v_x:

$$\frac{M_2^A}{M_0^A} = 1 + \frac{w_2 - w_0}{2w_1^A + w_2 + w_0} \cdot \frac{\gamma_x}{\gamma_1}$$

If a pure dipole–dipole interaction exists between two nuclei, W_2, W_1 and W_0 are in the ratio of $1 : 1/4 : 1/6$, and for $\gamma_A = \gamma_X$, the maximum signal amplification is 50%. This would be the case if the spin system consisted of two protons. For a spin system of type ^1H–^{13}C, the effect is four times as large (200%), since $\gamma_H / \gamma_C = 4$.

For the nuclear Overhauser effect, it is further important to note that the contribution of dipole–dipole interactions to the longitudinal relaxation time of two nuclei that are separated by a distance r is proportional to the factor $1/r^6$. A nuclear Overhauser effect, thus, can be observed only if the nuclei under consideration are in relatively close steric proximity, because only for this case the dipole–dipole interaction represents a significant relaxation mechanism. The correlation makes the NOE a useful tool for structure determination when it is necessary to decide which of two nuclei, A or B, is separated by a smaller distance from a third nucleus within the same molecule.

Finally, the intensity variations of individual lines in spin systems that are observed in connection with double resonance experiments are known as the general Overhauser effect. Hence, a very weak secondary field B_2 perturbs only those populations of energy levels that are linked through the irradiated line. For the appearance of this effect, the condition that $\gamma B_2^2 T_1 T_2 \approx 1$ should be met. The general Overhauser effect has found versatile and valuable applications in INDOR spectroscopy.

NMR of solids

To obtain an optimal NMR spectra, the analyte molecules should be tumbling rapidly relative to the magnet. The reason for this requirement is, in part, that the motions average the interactions with neighbouring molecules. In solids, however, the tumbling does not occur, and so the effects of the molecular environment do not average within the sample. As a result, the spectra are broad and nearly featureless and offer no chemical information.

Spectra of moderate resolution can be obtained by two ways. The first is to spin the sample at an angle of 54°44′, which is called magic angle, at a rate in excess of 200 kHz. This technique is called magic angle spinning (MAS). The second method is to average the magnetic interactions from the material surrounding the analyte molecules by using strong radio frequency fields. This is another instance of proton decoupling. For the solid, the method to remove the spin–spin interaction is called cross-polarization (CP).

REVIEW QUESTIONS

1. What is chemical shift? How is it measured relative to tetramethylsilane?

2. How is the applied magnetic field varied in practice?

3. Describe the factors which influence chemical shift.

4. What is spin–spin splitting?

5. Explain why NMR spectrum of benzene is observed at a lower field while that of acetylene is observed at a higher field strength.

6. Why is TMS used as a standard in NMR?

7. Define the terms: double resonance and spin tickling.

8. ^{13}C is NMR-active while ^{12}C is not. Explain.

9. Predict the number of signals and their multiplicities for the NMR spectrum of *p*-nitrotoluene.

10. Define coupling constant *J*.

11. What is meant by shielding and deshielding of a nucleus?

12. How are *cis* and *trans* stilbenes differentiated by NMR?

13. Acetylene protons are more shielded than ethylene protons. Account for this.

14. How many different types of protons are present in allyl bromide?

15. Calculate the chemical shift in ppm for a proton that has resonance at 126 Hz downfield from TMS on spectrometer that operates at 60 MHz.

16. How many hertz does 1 ppm correspond to for a PMR instrument operating at a radio frequency of 60 MHz?

17. Predict the low-and high-resolution NMR spectra of the following.

    i.   *n*-propane

    ii. ethyl methyl ether

    iii. *n*-propyl methyl ether

18. The low-resolution spectrum of $C_5H_{10}O_2$ exhibits three peaks at 3.2, 2.1 and 1.2 on delta scale with relative intensities 1:3:6 respectively. Predict the structure of the molecule.

19. Explain why PMR spectrum of *N*-methylacetamide show signals from two different *N*-methyl groups.

20. Sketch the first order PMR spectrum of each of the following compounds:

    i.   ethyl chloride

    ii. t-butylamine

    iii. 1, 1, 1 trifluoroethane

21. Explain nuclear Overhauser effect.

22. Predict the relative shapes of propane and 1-nitropropane.

Chapter 5

RAMAN SPECTROSCOPY

Introduction

Raman spectroscopy differs from rotational and vibrational spectroscopy in that it is concerned with the scattering of radiation by the sample, rather than with the absorption process. It was named after the Indian physicist C.V. Raman who first observed it in 1928. Both rotational and vibrational Raman spectroscopies are possible. The energy of the exciting radiation will determine which type of transition occurs—rotational transitions are lower in energy than vibrational transitions. In addition, rotational transitions are around three orders of magnitude slower than vibrational transitions. Therefore, collisions with other molecules may occur in the time in which the transition is occurring. A collision is likely to change the rotational state of the molecule, and so the definition of the spectrum obtained will be destroyed. Rotational spectroscopy is, therefore, carried out on gases at low pressures to ensure that the time between the collisions is greater than the time for a transition.

Scattering of Light by Molecules

Types of Scattering

Consider a light wave as a stream of photons each with energy $h\nu$. When each photon collides with a molecule, scattering of light can take place, two of which are considered below.

Elastic or Rayleigh scattering This occurs when the photon simply bounces off the molecule, with no exchange of energy. A vast majority of photons will scatter in this way (Figure 5.1)

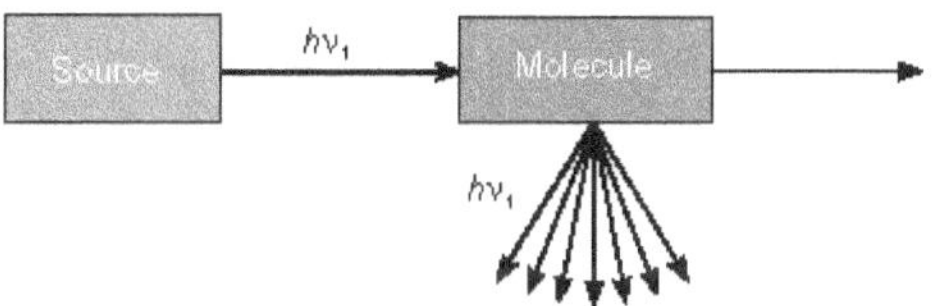

Figure 5.1 Rayleigh scattering by molecules

Inelastic or Raman scattering This occurs when there is an exchange of energy between the photon and the molecule, leading to the emission of another photon with a different frequency to the incident photon (Figure 5.2).

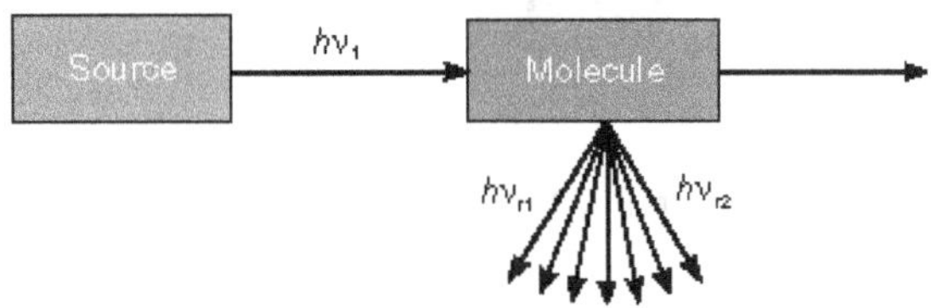

Figure 5.2   Raman scattering by molecules

Raman Effect

When light is scattered from a molecule, most of the photons are elastically scattered. The scattered photons have the same energy (frequency) and, therefore, wavelength as the incident photons. However, a small fraction of the light (approximately 1 in 10^7 photons) is scattered at optical frequencies different from, and usually lower than, the frequency of the incident photons. The process leading to this inelastic scatter is termed the **Raman effect**. Raman scattering can occur with a change in vibrational, rotational or electronic energy of a molecule. Chemists are concerned primarily with the vibrational Raman effect.

The Raman effect arises when a photon is incident on a molecule and interacts with the electric dipole of the molecule. It is a form of electronic (more accurately, vibronic) spectroscopy, although the spectrum contains vibrational frequencies. In classical terms, the interaction can be viewed as a perturbation of the molecule's electric field. In the quantum mechanics, the scattering is described as an excitation to a virtual state lower in energy than a real electronic transition with nearly coincident de-excitation and a change in vibrational energy. The scattering event occurs in 10^{-14} s or less. The virtual state description of scattering is shown in Figure 5.3a. Based on the frequency of the re-emitted photons, the scattering lines are grouped into two categories.

1. Stokes lines (re-emitted photons having slightly lower frequency and named after Sir George Gabriel Stokes who observed a similar phenomenon in fluorescence)

2. Anti-Stokes lines (re-emitted photons having higher frequency)

The difference in energy between the incident photon and the Raman-scattered photon is equal to the vibrational energy of the scattering molecule.

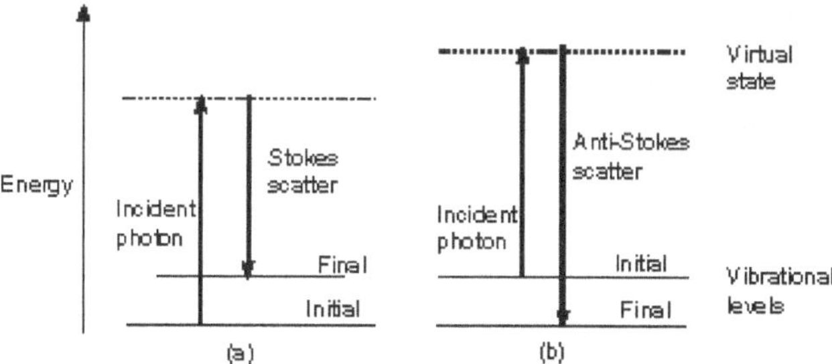

Figure 5.3   Energy level diagram for Raman scattering. (a) Stokes Raman scattering (b) Anti-Stokes Raman scattering. The population of the molecular energy levels follows Boltzmann distribution and hence Stokes lines are stronger (have more intensity) than anti-Stokes lines.

The energy difference between the incident and scattered photons is represented by the arrows of different lengths as shown in Figure 5.3a. Numerically, the energy difference between the initial and final vibrational levels, $\bar{V}$ or Raman shift in wave numbers (cm^{-1}), is calculted by

$$\bar{V} = \frac{1}{\lambda_{\text{incident}}} - \frac{1}{\lambda_{\text{scattered}}} \tag{5.1}$$

in which $\lambda_{\text{incident}}$ and $\lambda_{\text{scattered}}$ are the wavelengths (in cm) of the incident and Raman-scattered photons, respectively. The vibrational energy is ultimately dissipated as heat. Because of the low intensity of Raman scattering, the heat dissipation does not cause a measurable temperature rise in the material.

At room temperature, the thermal population of vibrational excited states is low, although not zero. Therefore, the initial state is the ground state, and the scattered photon will have lower energy (longer wavelength) than the exciting photon. This kind of Stokes-shifted scatter is usually observed in Raman spectroscopy. Figure 5.3a depicts Raman–Stokes scattering.

A small fraction of the molecules is in a vibrationally excited state. Raman scattering from vibrationally excited molecules leaves the molecule in the ground state. The scattered photon appears at higher energy, as shown in Figure 5.3b. This anti-Stokes-shifted Raman spectrum is always weaker than the Stokes-shifted spectrum, but at room temperature it is strong enough to be useful for vibrational frequencies less than about 1500 cm^{-1}. The Stokes and anti-Stokes spectra contain the same frequency information. The ratio of anti-Stokes to Stokes intensity at any vibrational frequency is a measure of temperature. Anti-Stokes Raman scattering is used for contactless thermometry. The anti-Stokes spectrum is also used when the Stokes spectrum is not directly observable, for example, because of poor detector response or spectrograph inefficiency.

A plot of Raman intensity versus Raman shift is a Raman spectrum.

The criteria for a molecule to be Raman-active are also different compared to other types of spectroscopy that require a permanent dipole moment, at least for diatomic molecules. A molecule will be Raman-active only if the following gross selection rule is fulfilled.

Selection Rules

Gross selection rule For spectroscopic techniques such as infrared spectroscopy, it is necessary for the molecule being analysed to have a permanent electric dipole. However, this is not the case for Raman spectroscopy; rather it is the polarizability (α) of the molecule which is important. The oscillating electric field of a photon causes the charged particles (electrons and, to a lesser extent, nuclei) in the molecule to oscillate. This leads to an induced electric dipole moment, μ_{ind}, which is given by

$$\mu_{ind} = \alpha\, E$$

This induced dipole moment then emits a photon, leading to either Raman or Rayleigh scattering. The energy of this interaction is also dependent on polarizability.

$$\text{Energy of interaction} = -1/2\alpha\, E^2$$

By comparing this equation with the equation for interaction energy for other forms of spectroscopy, it can be seen that the energies of Raman transitions are relatively weak. To counter this, a higher intensity of the exciting radiation is used.

For Raman scattering to occur, the polarizability of the molecule must vary with its orientation (Figure 5.4). One of the strengths of Raman spectroscopy is that this will be true for both heteronuclear and homonuclear diatomic molecules. Thus both homonuclear and heteronuclear diatomic molecules can be studied by Raman spectroscopy. Homonuclear diatomic molecules do not possess a permanent electric dipole, and so are undetectable by other methods such as infrared spectroscopy.

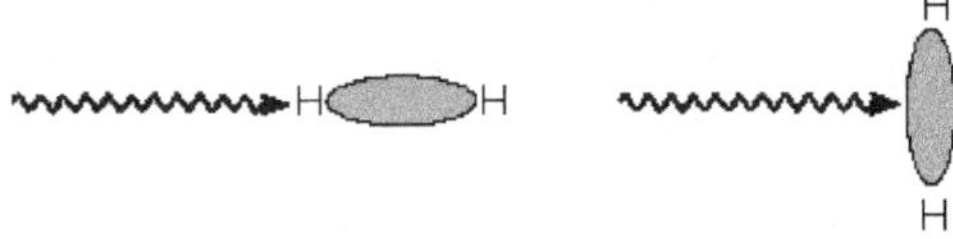

Figure 5.4 Polarizabilities of two homonuclear diatomic molecules

Selection rules arising from quantum mechanics A detailed discussion of the quantum mechanical basis for spectroscopy is beyond the scope of this brief chapter. Here, the quantum numbers and the allowed transitions involved are simply stated.

Rotational To find allowed rotational transitions, the total angular momentum quantum number (J) for rotational energy states must be considered. Raman spectroscopy involves a two-photon process, each of which obeys

$$\Delta J = \pm 1$$

Therefore, for the overall transition,

$$\Delta J = 0 \pm 2$$

where,

$\Delta J = 0$ corresponds to Rayleigh scattering,

$\Delta J = \pm 2$ corresponds to a Stokes trnsition, and

$\Delta J = -2$ corresponds to an anti-Stokes transition.

Vibrational   To find allowed vibrational transitions, the vibrational quantum number, (v), must be considered. For the overall transition,

$$\Delta v = \pm 1$$

Transitions are also possible for $\Delta v = \pm 2$ but are of weaker intensity.

Polarization Effects

Raman scatter is partially polarized, even for molecules in a gas or liquid, where the individual molecules are randomly oriented. The effect can be easily seen with an exciting source, which is plane polarized. In isotropic media, polarization arises because the induced electric dipole has components that vary spatially with respect to the coordinates of the molecule. Raman scatter from totally symmetric vibrations will be strongly polarized parallel to the plane of polarization of the incident light. The scattered intensity from non-totally symmetric vibrations is 3/4 as strong in the plane perpendicular to the plane of polarization of the incident light as in the plane parallel to it.

The situation is more complicated in a crystalline material. In that case, the orientation of the crystal is fixed in the optical system. The polarization components depend on the orientation of the crystal axes with respect to the plane of polarization of the input light, as well as on the relative polarization of the input and the observing polarizer.

Instrumentation

The basic set-up of a Raman spectrometer is shown in Figure 5.5. Note that the detector is orthogonal to the direction of the incident radiation, so as to observe only the scattered light. The source needs to provide intense monochromatic radiation, and so is usually a laser.

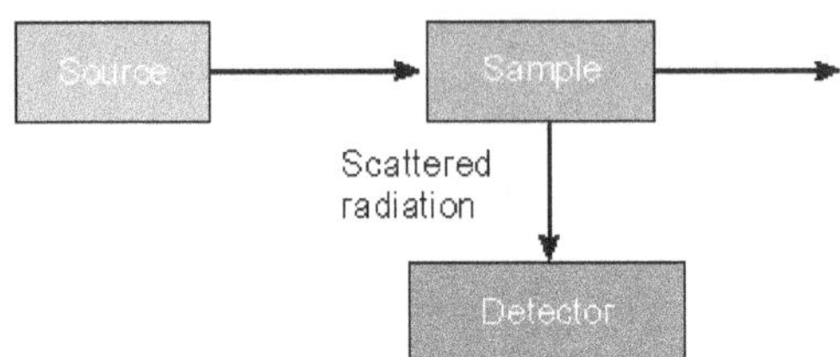

Figure 5.5   Block diagram of Raman spectrometer

A Raman system typically consists of four major components.

1.   Excitation source (laser)

2.   Sample illumination system and light collection optics

3.   Wavelength selector (filter or spectrophotometer)

4.   Detector (photodiode array, CCD or PMT)

A sample is normally illuminated with a laser beam in the ultraviolet (UV), visible (Vis) or near infrared (NIR) range. Scattered light is collected with a lens and is sent through interference filter or spectrophotometer to obtain Raman spectrum of the sample.

Since spontaneous Raman scattering is very weak, the main drawback of Raman spectroscopy is to differentiate it from the intense Rayleigh scattering. More precisely, the major problem here is not the Rayleigh scattering itself, but the fact that the intensity of stray light from Rayleigh scattering may greatly exceed the intensity of the useful Raman signal in close proximity to the laser wavelength. In many cases, the problem is resolved by simply cutting off the spectral range close to the laser line where the stray light has the most prominent effect. One can use the commercially available interference (notch) filter, which cuts off the spectral range of Â± 80–120 cm^{-1} from the laser line. This method is efficient in the elimination of stray light, but it does not allow the detection of low-frequency Raman modes in the range below 100 cm^{-1}.

Stray light is generated in the spectrometer mainly upon light dispersion on gratings, and it strongly depends on the grating quality. Raman spectrometers typically use holographic gratings that normally have much less manufacturing defects than the ruled ones. The stray light produced by the holographic gratings is about an order of magnitude less intense than from the ruled gratings of the same groove density.

Using multiple dispersion stages is another way of stray light reduction. Double and triple spectrometers allow taking Raman spectra without the use of notch filters. In such systems, Raman-active modes with frequencies as low as 3–5 cm^{-1} can be efficiently detected.

In earlier times, people primarily used single-point detectors such as photon-counting photomultiplier tubes (PMT). However, a single Raman spectrum obtained with a PMT detector in wavenumber scanning mode was taking substantial amount of time, slowing down any research or industrial activity based on Raman analytical technique. Nowadays, many researchers use multi-channel detectors like photodiode arrays (PDA) or, more commonly, a charge-coupled device (CCD) to detect the Raman-scattered light. The sensitivity and performance of modern CCD detectors are rapidly improving. In many cases, CCD is becoming the detector of choice for Raman spectroscopy.

Sample handling and illumination

As glass tubes are ideal for obtaining Raman spectra, sampling of liquids is much easier compared to IR spectroscopy. Water has a very weak Raman scattering, and hence vibrational spectra of aqueous solutions are almost always studied using Raman techniques. The small size of laser also makes it easy to investigate small solid samples, which generally require no container. For gases, however, the weakness of the Raman effect combined with low sample density makes the spectra hard to obtain without special care. It is almost essential to use a small cell with internal mirrors.

Selection of Laser Wavelength

The selection of laser wavelength is an important consideration for Raman spectroscopy. With modern equipment, several laser wavelengths may be employed so as to achieve the best detection of Raman signal.

For instance, many samples of "organic" or "biological" nature are quite fluorescent species. Exciting these samples (Figure 5.6) with a laser in green (532 nm) may promote fluorescence, and may swamp any underlying Raman spectrum to such an extent that it is no longer detectable. In this instance, the use of laser in red (633 nm) or NIR (785 nm) may provide a solution. With lower photon energy, a red or NIR laser may not promote electronic transition (and hence fluorescence), and so the Raman scatter may be far easier to detect.

Conversely, as one increases the wavelength, from green to red to NIR, the scattering efficiency will decrease, and so longer integration times or higher power lasers may be required.

Thus, it is often more practical to have a number of laser wavelengths to match the various sample properties one may encounter, be it resonance enhancement or penetration depth of fluorescence.

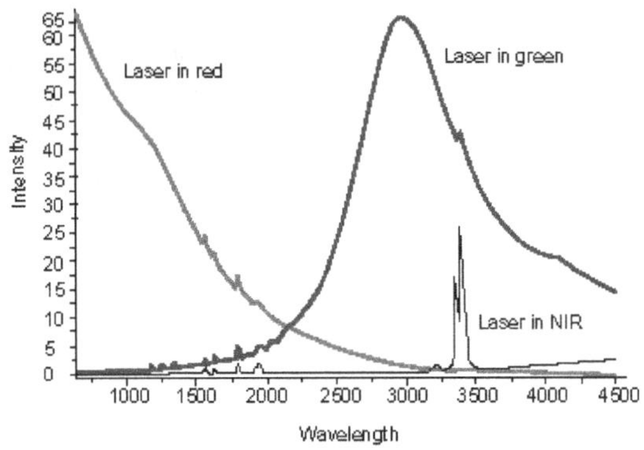

Figure 5.6   Green, red and NIR 785-nm laser excitation of a fluorescent sample The strong background seen with the green and red lasers swamps the Raman signal, whereas the 785-nm excitation is outside of the fluorescence range, enabling the Raman to be detected.

Ways to Improve Raman Signal Intensity

Raman signal is normally quite weak, and Raman spectroscopy techniques are being constantly improved. Many different ways of sample preparation, sample illumination or scattered light detection were introduced to enhance the intensity of Raman signal. Let us examine some of them here.

Stimulated raman scattering   This new non-linear phenomenon was observed in Raman signal when a sample was irradiated with a very strong laser pulse. In comparison with continuous wave (CW) lasers with electric field of about only 104 V cm^{-1}, pulsed lasers with electric field of about 109 V cm^{-1} transform a much larger portion of incident light into useful Raman scattering and substantially improve the signal-to-noise ratio.

Stimulated Raman scattering is an example of non-linear Raman spectroscopy. A very strong laser pulse with electric field strength >109 V cm^{-1} transforms up to 50% of all laser pulse energy into coherent beam at Stokes frequency $\nu_0-\nu_m$ (Figure 5.7). The Stokes beam is unidirectional with the incident laser beam. Only the mode which is strongest in the regular Raman spectrum, is greatly amplified. All other weaker Raman active modes are not present. The Stokes frequency is so strong that it acts as a secondary excitation source and generates the second Stokes line with frequency $\nu_0- 2\nu_m$. The second Stokes line generates the third one with the frequency $\nu_0-3\nu_m$, etc. Stimulated Raman technique enjoys four to five orders of magnitude enhancement of Raman signal as compared to the spontneous Raman scattering.

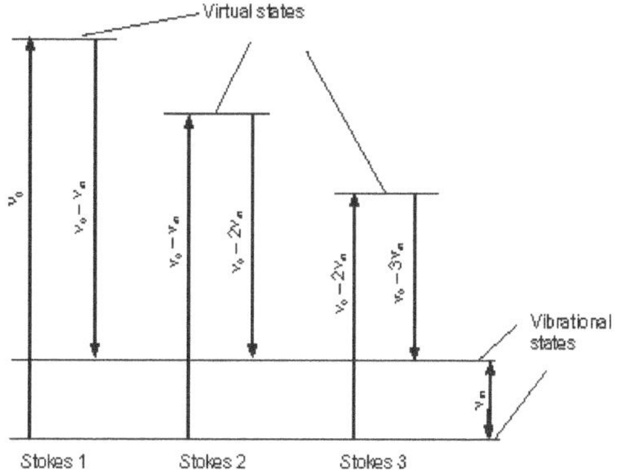

Figure 5.7   Stimulated Raman transitional schemes

Coherent anti-Stokes Raman (CARS)   Coherent anti-Stokes Raman (CARS) is another type of Raman spectroscopy. Unlike the traditional laser, two very strong collinear lasers irradiate a sample. The frequency of the first laser is usually constant, while the frequency of the second one can be tuned in a way that the frequency difference between the two lasers exactly equals the frequency of some Raman-active mode of interest. This particular mode will be the only strongest mode in Raman signal.

With CARS, we can obtain only one strong Raman peak of interest. In this case, a monochromator is not really required. A wideband interference filter and a detector behind the filter would do.

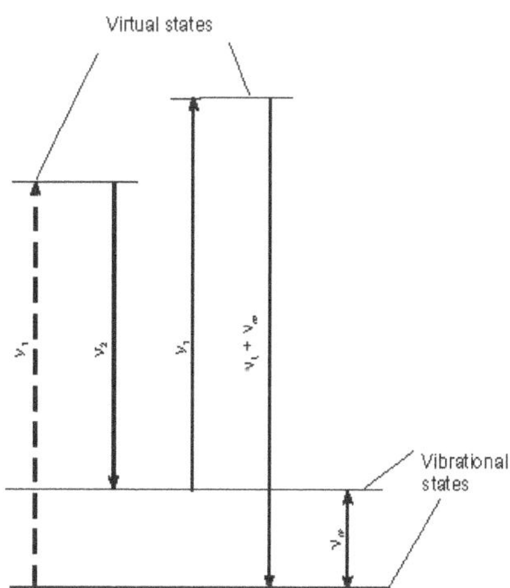

Figure 5.8   Transitional schemes for CARS

Two laser beams with frequencies v_1 and v_2 (where $v_1 > v_2$) interact coherently and, because of wave mixing, produce strong scattered light of frequency $v_2 - v_1$

(Figure 5.8). If the frequency difference between the two lasers is equal to the frequency ν_m of a Raman-active rotational, vibrational or any other mode, then a strong light of frequency $\nu_1 + \nu_m$ is emitted.

In other words, to obtain a strong Raman signal, the second laser frequency should be tuned in a way that $\nu_2 = \nu_1 - \nu_m$. Then the frequency of strong scattered light will be equal to $\nu_1 + \nu_m$, which is higher than the excitation frequency ν_1, and therefore considered to be anti-Stokes frequency.

Resonance Raman (RR) effect   Many substances, especially the coloured ones, absorb laser beam energy and generate strong fluorescence that contaminates Raman spectrum. This is one of the central problems in Raman spectroscopy, especially when UV lasers are used.

However, it was found that, under certain conditions, some types of coloured molecules can produce strong Raman scattering instead of fluorescence. This effect is called resonance Raman. The resonance Raman effect takes place when the excitation laser frequency is chosen in a way that it crosses the frequencies of electronic excited states and resonates with them. The intensity of Raman bands that originate from electronic transitions between these states is enhanced three to five orders of magnitude. Not all the bands of spontaneous Raman spectrum are enhanced. The so-called chromophoric group, which is responsible for molecule colouration, experiences the highest level of enhancement. The reason is that the chromophoric group normally has the highest level of light absorption.

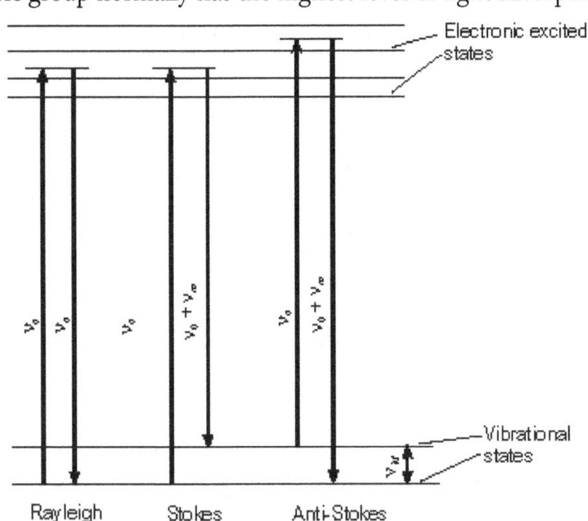

Figure 5.9   Resonance Raman transitional schemes

The highest intensity of resonance Raman signal is obtained when the laser frequency equals the first or second electronic excited state (Figure 5.9). Therefore, tunable lasers are the most appropriate choice for RR technique. Although the frequency of laser does not exactly match the desired electronic excited state, an impressive enhancement of Raman signal occurs.

SERS and SERRS

In SERS (surface-enhanced Raman spectroscopy), the Raman signal from molecules adsorbed on certain metal surfaces can be five to six orders of magnitude stronger than the Raman signal from the same molecules in bulk volume. The exact reason for such dramatic improvement is still unknown. However, since the intensity of Raman signal is proportional to the square of electric dipole moment $\rho = \alpha E$, there are two possible reasons: the enhancement of polarizability α, or the enhancement of electrical field E.

The enhancement of polarizability may occur because of a charge-transfer effect or a chemical bond formation between the metal surface and the molecules under observation. This is called chemical enhancement.

The enhancement of E takes into account the interaction of laser beam with the irregularities on the metal surface, such as metal microparticles or roughness profile. It is believed that laser light excites conduction electrons at the metal surface, leading to surface plasma resonance and strong enhancement of electric field E. This is also called electromagnetic enhancement.

In all cases, the choice of appropriate surface substrate is very important. The most popular and universal substrates used for SERS are electrochemically etched silver electrodes as well as silver and gold colloids with average particle size below 20 nm.

One disadvantage of SERS is the difficulty of spectra interpretation. The signal enhancement is so dramatic that the Raman bands that are very weak and unnoticeable in spontaneous Raman spectra can appear in SERS. Some trace contaminants can also contribute additional peaks. On the other hand, because of chemical interactions with the metal surface, certain peaks that are strong in conventional Raman might not be present in SERS at all. The non-linear character of signal intensity as a function of concentration complicates the things further. Careful consideration of all physical and chemical factors should be done while interpreting SERS spectra, which makes it extremely difficult for practical use.

The SERRS (surface-enhanced resonance Raman spectroscopy) technique was developed to overcome such complications as above. It utilizes both surface enhancement effect and Raman resonance effect to enhance the Raman signal intensity. The main advantage of SERRS is that its spectra very much resemble regular resonance Raman spectra, which makes it much easier to interpret.

Resonance-Enhanced Raman Scattering

Raman spectroscopy is conventionally performed with green, red or near-infrared lasers. The wavelengths are below the first electronic transitions of most molecules, as assumed by scattering theory. The situation changes if the wavelength of the exciting laser is within the electronic spectrum of a molecule. In that case, the intensity of some Raman-active vibrations increases by a factor of 10^2–10^4. This resonance enhancement or resonance Raman effect can be quite useful.

Metalloporphyrins, carotenoids and several other classes of biologically important molecules have strongly allowed electronic transitions in the visible region. The spectrum of chromophoric moiety is resonance-enhanced and that of the surrounding protein matrix is not. This allows the physical biochemists to probe the chromophoric site (often the active site) without spectral interference from the surrounding protein. Resonance Raman spectroscopy is also a major probe of the chemistry of fullerenes, polydiacetylenes and other 'exotic' molecules that strongly absorb in the visible region. Although many more molecules absorb in the ultraviolet, the high cost of lasers and optics for this spectral region have limited the UV resonance Raman spectroscopy to a small number of specialists.

The vibrations whose Raman bands are resonance-enhanced fall into two or three general classes. The most common case is Franck–Condon enhancement in which a component of the normal coordinate of vibration is in a direction in which the molecule expands during an electronic excitation. The more the molecule expands along this axis when it absorbs light, the larger the enhancement factor. The easily visualized ring-breathing (in-plane expansion) modes of porphyrins fall into this class. Vibrations that couple two electronic excited states are also resonance-enhanced. This mechanism is called vibronic enhancement. In both cases, enhancement factors roughly follow the intensities of absorption spectrum.

Resonance enhancement does not begin at a sharply defined wavelength. In fact, enhancement of 5 to 10 times is commonly observed if the exciting laser is even within a few hundred wave numbers below the electronic transition of a molecule. This pre-resonance enhancement can be experimentally useful.

Surface-Enhanced Raman Scattering

The Raman scattering from a compound (or ion) adsorbed on or even within a few Angstroms of a structured metal surface can be 10^3–10^6X greater than in solution. The strongest surface-enhanced Raman scattering is observed on silver, but is observable on gold and copper as well. At practical excitation wavelengths, enhancement on other metals is unimportant. SERS arises from two mechanisms.

The first is an enhanced electromagnetic field produced at the surface of the metal. When the wavelength of the incident light is close to the plasma wavelength of the metal, conduction electrons on the metal surface are excited into an extended surface electronic excited state called surface plasmon resonance. Molecules adsorbed or in close proximity to the surface, experience an exceptionally large electromagnetic field. Vibrational modes normal to the surface are the most strongly enhanced.

The second mode of enhancement is by the formation of a charge-transfer complex between the surface and analyte molecule. The electronic transitions of many charge-transfer complexes are in the visible, so that resonance enhancement occurs.

Molecules with lone pair of electrons or π clouds show the strongest SERS. The effect was first discovered with pyridine. Other aromatic nitrogen- or oxygen-containing compounds such as aromatic amines or phenols are strongly SERS-active. This effect can also be seen with other electron-rich functionalities, such as carboxylic acids.

The intensity of surface plasmon resonance is dependent on many factors including the wavelength of the incident light and the morphology of the metal surface. The wavelength should match the plasma wavelength of the metal. This is about 382 nm for a 5-mm silver particle, but can be as high as 600 nm for larger ellipsoidal silver particles. The plasma wavelength is 650 nm for copper and gold; the other two metals show SERS at wavelengths in the 350–1000 nm region. The best morphology for surface plasmon resonance excitation is a small (<100 nm) particle or an atomically rough surface.

SERS is used to study the monolayers of materials adsorbed on metals, including electrodes. Many formats other than electrodes can be used. The most popular materials include colloids, metal films on dielectric substrates and, recently, arrays of metal particles bound to metal or dielectric colloids through short linkages. Although SERS allows an easy observation of Raman spectra from solution concentrations in the micromolar (1×10^{-6}) range, the slow adsorption kinetics and competitive adsorption of materials limit its application in analytical chemistry.

Diagnostic Structural Analysis

Raman Spectroscopy for Analysis and Monitoring

The Raman scattering technique is a vibrational molecular spectroscopic technique which is derived from an inelastic light scattering process. With Raman spectroscopy, a laser photon is scattered by a sample molecule and it loses (or gains) energy during the process. The amount of energy lost is seen as a change in energy (wavelength) of the irradiating photon. The energy loss is characteristic for a particular bond in the molecule. Raman spectroscopy can best be thought of as producing a precise spectral fingerprint which is unique to a molecule or indeed an individual molecular structure. In this respect, it is similar to the more commonly found FT-IR spectroscopy. However, unlike FT-IR, there are a distinct number of advantages in Raman spectroscopy.

* It can be used to analyse aqueous solutions since it does not suffer from the large water absorption effects that are found with FT techniques.

Table 5.1   Vibration frequencies of funtional groups in Ramex nd infrared region

Functional group/vibration	Region	Raman	Infrared
Lattice vibrations in crystals, LA modes	$10–200$ cm^{-1}	Strong	Strong
δ (CC) aliphatic chains	$250–400$ cm^{-1}	Strong	Weak
v (Se—Se)	$290–330$ cm^{-1}	Strong	Weak
v (S—S)	$430–550$ cm^{-1}	Strong	Weak
v (Si—O—Si)	$450–550$ cm^{-1}	Strong	Weak
v (Xmetal—O)	$150–450$ cm^{-1}	Strong	Medium-weak
v (C—I)	$480–660$ cm^{-1}	Strong	Weak
v (C—Br)	$500–700$ cm^{-1}	Strong	Weak
v (C—Cl)	$550–800$ cm^{-1}	Strong	Weak
v (C—S) aliphatic	$630–790$ cm^{-1}	Strong	Medium
v (C—S) aromatic	$1080–1100$ cm^{-1}	Strong	Medium
v (O—O)	$845–900$ cm^{-1}	Strong	Weak
v (C—O—C)	$800–970$ cm^{-1}	Medium	Weak
v (C—O—C) asym	$1060–1150$ cm^{-1}	Weak	Strong
v (CC) alicyclic, aliphatic chain vibrations	$600–1300$ cm^{-1}	Medium	Medium
v	$1000–1250$ cm^{-1}	Strong	Weak
v (CC) aromatic ring chain vibrations	*$1580, 1600$ cm^{-1}	Strong	Medium
	*$1450, 1500$ cm^{-1}	Medium	Medium
	*1000 cm^{-1}	Strong/medium	Weak
δ (CH$_3$)	1380 cm^{-1}	Medium	Strong
δ (CH$_2$)	$1400–1470$ cm^{-1}	Medium	Medium
δ (CH$_3$)asym			
δ (CH$_2$)	$1400–1470$ cm^{-1}	Medium	Medium
δ (CH$_3$) asym	$1340–1380$ cm^{-1}	Strong	Medium
v (C—NO$_2$)) asym	$1530–1590$ cm^{-1}	Medium	Strong

Table 5.1 (Continued)

Functional group/vibration	Region	Raman	Infrared
ν(N N) aromatic	1410–1440 cm^{-1}	Medium	-
ν (N N) aliphatic	1550–1580 cm^{-1}	Medium	-
δ(H$_2$O)	–1640 cm^{-1}	Weak broad	Strong
ν (C N)	1610–1680 cm^{-1}	Strong	Medium
ν(C C)	1500–1900 cm^{-1}	Strong	Weak
ν(C O)	1680–1820 cm^{-1}	Medium	Strong
ν(C C)	2100–2250 cm^{-1}	Strong	Weak
ν(C N)	2220–2255 cm^{-1}	Medium	Strong
ν(—S—H)	2550–2600 cm^{-1}	Strong	Weak
ν(C—H)	2800–3000 cm^{-1}	Strong	Strong
ν((C—H))	3000–3100 cm^{-1}	Strong	Medium
ν((C—H))	3300 cm^{-1}	Weak	Strong
ν(N—H)	3300–3500 cm^{-1}	Medium	Medium
ν(O—H)	3100–3650 cm^{-1}	Weak	Strong

- The intensity of spectral features in a solution is directly proportional to the concentration of the particular species.

- Raman spectra are generally robust to temperature changes.

- It requires little or no sample preparation. It does not need the use of Nujol, or KBr matrices and is largely unaffected by sample cell materials such as glass.

- The use of Raman microscope such as the LabRAM provides very high level of spatial resolution and depth discrimination, which are not found with the FT methods of analysis.

These advantages along with its highly specific nature mean that Raman spectroscopy has become a very powerful tool for analysis and chemical monitoring (Table 5.1). Depending upon instrumentation, it is a technique which can be used for the analysis of solids, liquids and solutions and can even provide information on physical characteristics such as crystalline phase and orientation, polymorphic forms and intrinsic stress.

Applications

1. From the study of Raman spectra of crystalline and amorphous state of a substance, it has been observed that the amorphous state gives rise to broad and diffuse bands while the crystalline state gives fine sharp lines.

2. Ionic equilibria in a solution can also be studied. For example,

$$HNO_3 + H_2O \leftrightarrow H_3O^+ + NO_3^-$$

   It is possible to calculate the dissociation constant of nitric acid by monitoring the intensity of the nitrate ion in nitric acid.

3. Carbon monoxide gives a Raman line at 2155 cm^{-1} closely similar to —C N— at 2200 cm^{-1}. This shows that, like cyanides, carbon monoxide has a triple bond.

4. Hydrogen cyanide shows two lines at 2062 and 2094 cm^{-1}. This may be due to the two isomers HCN and HNC in dynamic equilibrium.

   There must be a change in polarizability during vibration for that vibration to inelastically scatter radiation.

Examples of Raman-Active and Inactive Vibrations in CO$_2$

Symmetric stretch 1340 cm^{-1} Asymmetric stretch 2350 cm^{-1}

O=C=O   O=C=O

Thus carbon dioxide has a strong band in its Raman spectrum at 1340 cm^{-1} and two strong bands in its infrared spectrum at 668 and 2349 cm^{-1}. As none of these occurs both in the Raman and infrared spectrum, CO$_2$ has a centre of symmetry in accordance with the law of mutual exclusion. It states that for molecules with a centre of symmetry, transitions that are allowed in the infrared are forbidden in the Raman spectra and vice versa.

Applications of Resonance Raman Spectroscopy

1. Resonance Raman spectroscopy is well suited to the study of biostructural problems and complex biological materials. The electronic absorption is often localized in a particular group of a large molecule, the so-called chromophore. The RR enhancement permits the isolation of Raman bands of the chromophore and the structural units located close to it, and therefore gives the information about sites of a molecule, which makes assignment of a complicated molecule possible. It limits the amount of structural information obtainable from such a spectrum, but it represents a great deal of simplification, which is focussed directly on a specific part of the molecule. The resonance

Raman spectra obtained by using different lasers of wavelength coincident with different bands of electronic spectrum of the species may be more informative than a normal Raman spectrum. Solute and solvent bands and relatively higher concentrations of other reagents, such as buffers, do not interfere. This is particularly advantageous when very large molecules such as biomolecules are studied.

2. Certain transitions normally forbidden in conventional Raman scattering are allowed in resonance Raman scattering. It provides information about the molecular system under investigation.

3. It has a great potential for the study of free radicals and transient species. The derived information will be complementary to the results from electron spin resonance (ESR) spectra. The resonance Raman spectroscopy is not affected by line broadening and other interferences that limit the applicability of ESR spectroscopy.

4. It offers a good method for the determination of anharmonicity constants of resonance-enhanced vibrational modes. Since the harmonic frequency and anharmonicities can be accurately calculated from the progression of overtone bands, interactions of solvent with scattering solute molecules can be detected even at very low concentrations.

5. Since the frequency of Raman vibrations depends on the electronic ground state, whereas the scattered Raman intensity is dependent on the absolute location of the incident laser energy to the electronic absorption, information about properties of a molecule in an electronic excited state can be obtained from RRS. The measurement of scattered Raman intensity as a function of the incident laser energy produces the so-called excitation profile (EP) from which the strength of interaction between the electronic excited state and the vibrational mode can be estimated.

6. Detailed studies of excitation profiles of Raman bands are likely to demonstrate the close relationship between Raman and electronic absorption spectroscopy. Resonance Raman spectra can sometimes help in the assignment of electronic transitions. The availability of a range of excitation frequencies from laser sources and the development of tunable lasers have resulted in a growing exploitation of resonance Raman spectroscopy.

Raman Microspectroscopy

Raman spectroscopy offers several advantages for microscopic analysis. Since it is a scattering technique, specimens do not need to be fixed or sectioned. Raman spectra can be collected from a very small volume (< 1 μm in diameter); these spectra allow the identification of species present in that volume. Water does not interfere very strongly. Thus, Raman spectroscopy is suitable for microscopic examination of minerals or materials such as polymers and ceramics, cells and proteins. The Raman microscope begins with a standard optical microscope, and adds an excitation laser, a monochromator and a sensitive detector (such as a CCD or PMT). FT-Raman has also been used with microscopes.

In **direct imaging**, the whole field of view is examined for scattering over a small range of wavenumbers (Raman shifts). For instance, the wavenumber characteristic for cholesterol could be used to record the distribution of cholesterol within a cell culture.

The other approach is **hyperspectral imaging**, in which thousands of Raman spectra are acquired from all over the field of view. The data can then be used to generate images showing the location and amount of different components. Taking the cell culture example, a hyperspectral image could show the distribution of cholesterol, as well as proteins, nucleic acids and fatty acids. Sophisticated signal- and image-processing techniques can be used to ignore the presence of water, culture media, buffers and other interferents.

Raman microscopy, and in particular confocal microscopy, has very high spatial resolution. For example, the lateral and depth resolutions were 250 nm and 1.7 μm, respectively, using a confocal Raman microspectrometer with the 632.8-nm line from a He–Ne laser with a pinhole of 100 μm diameter.

Since the objective lenses of microscopes focus the laser beam to several microns in diameter, the resulting photon flux is much higher than that achieved in conventional Raman set-ups. This has the added advantage of enhanced fluorescence quenching. However, the high photon flux can also cause sample degradation, and for this reason some set-ups require a thermally conducting substrate (which acts as a heat sink) in order to mitigate this process.

By using Raman microspectroscopy, *in vivo* time- and space-resolved Raman spectra of microscopic regions of samples can be measured. As a result, the fluorescence of water, media and buffers can be removed. Consequently, *in vivo* time- and space-resolved Raman spectroscopy is suitable to measure cells, proteins, organs and erythrocytes.

Raman microscopy for biological and medical specimens generally uses NIR lasers (785-nm diodes and 1064-nm Nd:YAG are common). This reduces the risk of damaging the specimen by applying high power However, the intensity of NIR Raman is low (owing to the ω^{-4} dependence of Raman scattering intensity), and most detectors require very long collection times. Recently, more sensitive detectors have become available, making the technique better suited to general use. Raman microscopy of inorganic specimens, such as rocks, ceramics and polymers, can use a broader range of excitation wavelengths.

REVIEW QUESTIONS

1. Why is mercury arc suitable for studying Raman effect?

2. What are Stokes and anti-Stokes lines?

3. What is Raman effect?

4. Explain the rule of mutual exclusion.

5. Discuss the mechanism of Raman effect.

6. What are the limitations of Raman spectroscopy?

7. Discuss some important applications of Raman spectroscopy.

8. Mention the selection rules for Raman spectroscopy.

9. Using a block diagram, explain the instrumentation of Raman spectrometer.

10. How is stimulated Raman scattering produced?

11. Write a brief note on CARS.

12. Explain in detail, surface-enhanced Raman scattering.

13. Explain the use of lasers in Raman spectroscopy.

14. What is resonance Raman spectroscopy? Explain.

Chapter 6

MASS SPECTROMETRY

Introduction

Mass spectrometry is a method of production of gas-phase ions from a sample, and their resulting separation according to their mass-to-charge ratio. In this method, a substance is bombarded with an electron beam of sufficient energy to fragment the molecule. The positive fragments (cations and radical cations) produced, are accelerated in vacuum through a magnetic field and are sorted on the basis of mass-to-charge ratio. Since the bulk of ions produced in a mass spectrometer carry a unit positive charge, the value m/z is equivalent to the molecular weight of the fragment. The analysis of mass spectroscopy information involves re-assembling of fragments and working backwards to generate the original molecule.

Mass spectrometry is a microanalytical technique in which only a few nanomoles of the sample is required to obtain characteristic information with respect to the structure and molecular weight of the analyte. It is a destructive technique, which means that the sample is consumed during analysis.

In most cases, the nascent molecular ion of the analyte produces fragment ions and the resulting fragmentation pattern gives rise to the mass spectrum. Thus, the mass spectrum of each compound is unique and can be used as a chemical fingerprint to characterize the sample.

Principle

The physics behind mass spectrometry is that a charged particle passing through a magnetic field is deflected along a circular path on a radius that is proportional to the mass-to-charge ratio, i.e., m/z. In an **electron impact** mass spectrometer, a high-energy beam of electrons is used to displace an electron from the organic molecule to form a **radical cation** known as the **molecular ion.** If the molecular ion is too unstable, then it can **fragment** to give other smaller ions (Figure 6.1). The ions are collected, focused into a beam, accelerated into the magnetic field, and deflected along circular paths according to the masses of ions. By adjusting the magnetic field, the ions can be focused on the detector and recorded.

The deflection of the ion depends on its mass, charge and velocity. For a given charge, velocity and deflecting force, the deflection for a heavy particle is less as compared to that of a light ion. Thus, a number of lines, each containing ions with the same m/z values, are obtained.

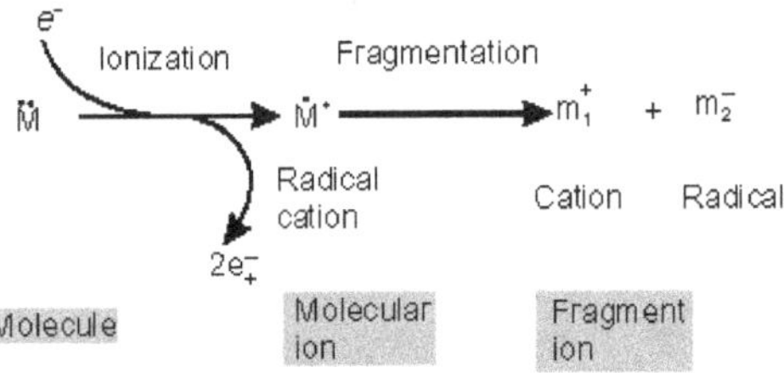

Figure 6.1   Electron impact mass spectrometer

MASS SPECTROMETER

A very low concentration of sample molecules is allowed to leak into the ionization chamber (which is under very high vacuum), where they are bombarded by a high-energy electron beam. The fragment molecules and the positive ions produced are accelerated through a charged array into an analysing tube. The path of the charged molecules is bent by an applied magnetic field. Ions having low mass (low momentum) will be deflected most by this field and will collide with the walls of the analyser. Likewise, high-momentum ions will not be deflected enough and will also collide with the analyser wall. Ions having the proper mass-to-charge ratio, however, will follow the path of the analyser, exit through the slit and collide with the collector (Figure 6.2). This generates an electric current, which is then amplified and detected. By varying the strength of the magnetic field, the mass-to-charge ratio can be continuously varied

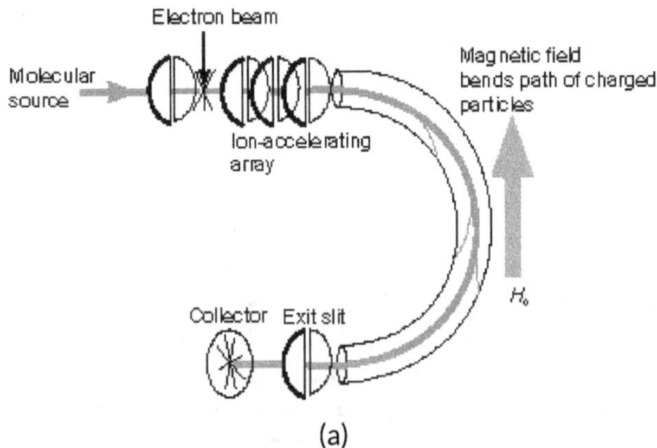

Figure 6.2   Mass spectrometer (*Continues*

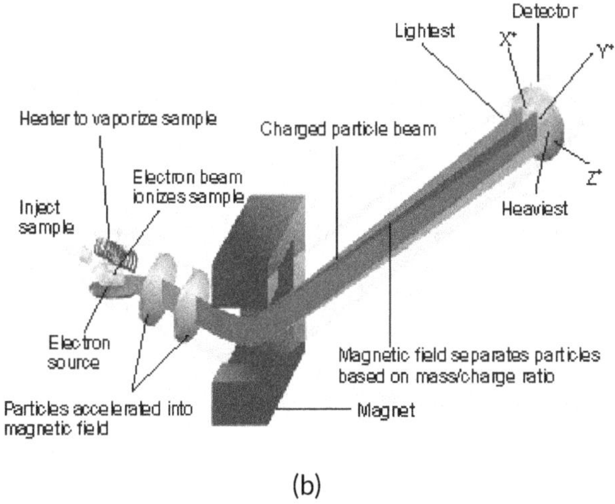

(b)

Figure 6.2　Mass spectrometer

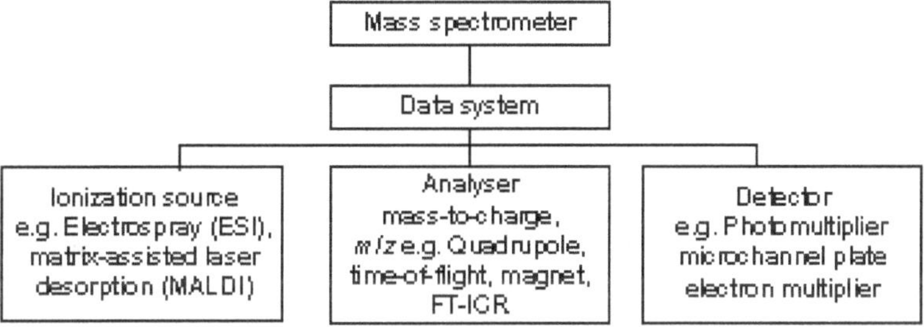

Figure 6.3　Simplified schematic representation of a mass spectrometer

Instrumentation

A mass spectrometer principally consists of three distinct regions: (1) ionizer (2) ion analyser (3) detector.

Sample Introduction (Ionization Source)

The selection of a sample inlet depends upon the sample and the sample matrix. Most ionization techniques are designed for gas-phase molecules; hence the inlet must transfer the analyte into the source as a gas-phase molecule. If the analyte is sufficiently volatile and thermally stable, a variety of inlets are available. Gases and samples with high vapour pressure are introduced directly into the source region. Liquids and solids are usually heated to increase their vapour pressure for analysis. If the analyte is thermally labile (it decomposes at high temperatures) or if it does not have sufficient vapour pressure, the sample must be directly ionized from the condensed phase. These direct ionization techniques require special instrumentation and are more difficult to use. However, they greatly extend the range of compounds that may be analysed by mass spectrometry. Commercial instruments are available that use direct ionization techniques to routinely analyse proteins and polymers with molecular weights greater than 1,00,000 daltons.

Direct vapour inlet   The simplest method for sample introduction is the direct vapour inlet. The gas phase analyte is introduced directly into the source region of the mass spectrometer through a needle valve. Pump-out lines are usually included to remove air from the sample. This inlet works well for gases, liquids, or solids with a high vapour pressure. Samples with low vapour pressure are heated to increase their vapour pressure. Since this inlet is limited to stable compounds and modest temperatures, it only works for some samples.

Gas chromatography   Gas chromatography is probably the most common technique for introducing samples into a mass spectrometer. Complex mixtures are routinely separated by gas chromatography, and mass spectrometry is used to identify and quantitate individual components. Several different interface designs are used to connect these two instruments. The most significant characteristics of the inlets are the amount of GC carrier gas that enters the mass spectrometer and the amount of analyte that enters the mass spectrometer. If a large flow of GC carrier gas enters the mass spectrometer, it will increase the pressure in the source region. Maintaining the required source pressure will require larger and more expensive vacuum pumps. The amount of analyte that enters the mass spectrometer is important for improving the detection limits of the instrument. Ideally, all the analyte and none of the GC carrier gases would enter the source region.

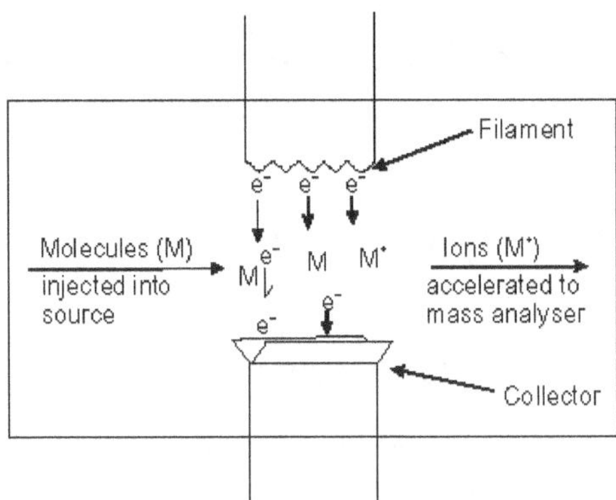

Figure 6.4   Electron ionization source

The most common GC/MS interface now used is a capillary GC column. Since the carrier gas flow rate is very small for these columns, the end of the capillary is inserted directly into the source region of the mass spectrometer. The entire flow from the GC enters the mass spectrometer. Since capillary columns are now very common, this inlet is widely used. The electrons used for ionization are produced by passing current through a wire filament (Figure 6.4). The amount of current controls the number of electrons emitted by the filament. An electric field accelerates these electrons across the source region to produce a beam of high-energy electrons. When an analyte molecule passes through this electron beam, a valence shell electron can be removed from the molecule to produce an ion.

Ionizer

A variety of ionization techniques are used in mass spectrometry. Most ionization techniques excite the neutral analyte molecule, which then ejects an electron to form a radical cation $(M_\bullet^+)$. Other ionization techniques involve ion–molecule reactions that produce adduct ± .ions (MH). The most important considerations are the physical state of the analyte and the + ** ionization energy. Electron ionization and chemical ionization are only suitable for gas-phase ionization. Fast atom bombardment, secondary ion mass spectrometry, electrospray, and matrix-assisted laser desorption are used to ionize condensed phase samples. The ionization energy is significant because it controls the amount of fragmentation observed in the mass spectrum.

Although this fragmentation complicates the mass spectrum, it provides structural information for the identification of unknown compounds. Some ionization techniques are very soft and only produce molecular ions; other techniques are very energetic and cause ions to undergo extensive fragmentation. Although this fragmentation complicates the mass spectrum, it provides structural information for the identification of unknown compounds.

Electron ionization   Electron ionization (EI) is the most common ionization technique used for mass spectrometry. EI works well for many gas-phase molecules, but it does have some limitations. Although the mass spectra are very reproducible and are widely used for spectral libraries, EI causes extensive fragmentation so that the molecular ion is not observed for many compounds. Fragmentation is useful because it provides structural information for interpreting unknown spectra.

In gas chromatography–mass spectrometry (GC–MS), the charged particles (ions) required for mass analysis are formed by electron impact (EI) ionization. The gas molecules exiting the GC are bombarded by a high-energy electron beam (70 eV). An electron that strikes a molecule may impart enough energy to remove another electron from that molecule. Methanol, for example, would undergo the following reaction in the ionizing region:

$$CH_3OH + 1\,electron \longrightarrow CH_3OH^{+\cdot} + 2\,electrons$$

EI usually produces singly charged ions containing one unpaired electron. A charged molecule that remains intact is called the molecular ion. Energy imparted by the electron impact and, more important, instability in a molecular ion can cause that ion to break into smaller pieces (fragments). The methanol ion may fragment in various ways, with one fragment carrying the charge and one fragment remaining uncharged. For example,

$$CH_3OH^+ \text{ (molecular ion)} \rightarrow CH_2OH^+ \text{ (fragment ion)} + H$$

$$\text{(or)}$$

$$CH_3OH^+ \text{ (molecular ion)} \rightarrow CH_3 + \text{ (fragment ion)} + OH$$

Chemical ionization   Chemical ionization (CI) is a 'soft' ionization technique that produces ions with little excess energy. As a result, less fragmentation is observed in the mass spectrum. Since this increases the abundance of the molecular ion, the technique is complementary to 70 eV EI. CI is often used to verify the molecular mass of an unknown.

Only slight modifications of an EI source region are required for CI experiments.

In chemical ionization, the source is enclosed in a small cell with openings for the electron beam, the reagent gas and the sample. The reagent gas is added to this cell at approximately 10 Pa (0.1 torr) pressure. This is higher than the 10 Pa (10 torr) pressure typical for a mass spectrometer source. At 10 Pa, the mean free path between collisions is approximately 2 metres, and ion–molecule reactions are unlikely. In the CI source, however, the mean free path between collisions is only 10 metres, and analyte molecules undergo many collisions with the reagent gas. The reagent gas in the CI source is ionized with an electron beam to produce a cloud of ions. The reagent gas ions in this cloud react and produce adduct ions like CH_5^+, which are excellent proton donors.

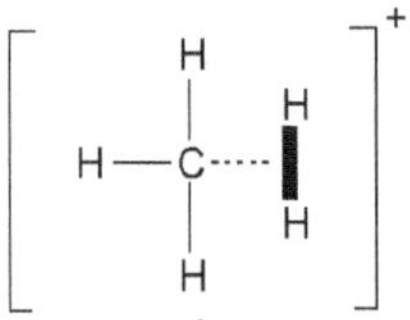

$$CH_5^+ \text{ ion}$$

When analyte molecules (M) are introduced to a source region with this cloud of ions, the reagent gas ions donate a proton to the analyte molecule and produce MH ions. The energetics of the proton transfer is controlled by using different reagent gases. The most common reagent gases are methane, isobutane and ammonia. Methane is the strongest proton donor commonly used with a proton affinity (PA) of 5.7 eV. For softer ionization, isobutane (PA 8.5 eV) and ammonia (PA 9.0 eV) are frequently used. Acid–base chemistry is frequently used to describe the chemical ionization reactions. The reagent gas must be a strong Bronsted acid to transfer a proton to the analyte. Fragmentation is minimized in CI by reducing the amount of excess energy produced by the reaction. Because adduct ions have little excess energy and are relatively stable, CI is very useful for molecular mass determination. Some typical reactions in a CI source are shown in Table 6.1.

Table 6.1   Chemical ionization reactions

Reaction of reagent gas	Ionization reaction
Reagent gas reacts to form ions	$CH_4 + e^- \rightarrow CH_4 + 2e^-$
Reagent gas to form adducts	$CH_4^+ + CH_4 \rightarrow CH_3 + CH_5^+$ (or) $CH_4 \rightarrow CH_3 + H^1$ $CH_3 + CH_4 \rightarrow C_2H_5 + H_2$
Reagent gas to form adducts	$CH_5^+ + m \rightarrow CH_4 + mH^+$ $C_2H_5^+ + M \rightarrow C_2H_4 + MH^+$ $CH_3^+ + M \rightarrow CH_4 + (M-H)^+$

Fast atom bombardment and secondary ion mass spectrometry   Fast atom bombardment (FAB) and secondary ion mass spectrometry (SIMS) both use high-energy atoms to sputter and ionize the sample in a single step. In these techniques, a beam of rare gas neutrals (FAB) or ions (SIMS) is focussed on the liquid or solid sample. The impact of this high-energy beam causes the analyte molecules to sputter into the gas-phase and ionize in a single step (Figure 6.5). The exact mechanism of this process is not well understood, but these techniques work well for compounds with molecular weights up to a few thousand daltons. Since no heating is required, sputtering techniques (especially FAB) are useful for studying thermally labile compounds that decompose in conventional inlets.

Atmospheric pressure ionization and electrospray ionization   Atmospheric pressure ionization (API) sources ionize the sample at atmospheric pressure and then transfer the ions into the mass spectrometer. These techniques are used to ionize thermally labile samples such as peptides, proteins and polymers directly from the condensed phase. The sample is dissolved in an appropriate solvent and this solution is introduced into the mass spectrometer. With conventional inlets, the solvent increases the pressure in the source region of the mass spectrometer. In addition, Joule–Thompson cooling of the liquid, as it enters the vacuum, causes the solvent droplets to freeze. The frozen clusters trap the analyte molecules and reduce the sensitivity of the expetiment.

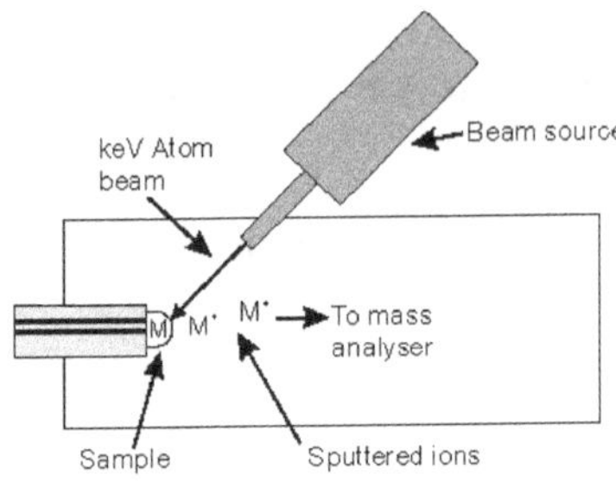

Figure 6.5 Fast atom bombardment source

API sources introduce the sample through a series of differentially pumped stages. This maintains the large pressure difference between the ion source and the mass spectrometer (Figure 6.6) without using extremely large vacuum pumps. In addition, a drying gas is used to break up the clusters that form as the solvent evaporates. Because the analyte molecules have more momentum than the solvent and air molecules, they travel through the pumping stages to the mass analyser.

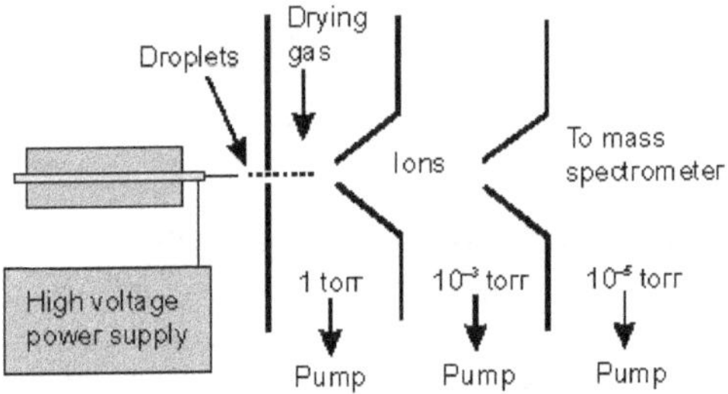

Figure 6.6 Electrospray ionization source

Electrospray ionization (ESI) is the most common API application. It has undergone remarkable progress in the recent years and is frequently used for LC/MS of thermally labile and high-molecular-weight compounds. The electrospray is created by applying a large potential between the metal inlet needle and the first skimmer in an API source (Figure 6.6). The mechanism for the ionization process is not well understood, and there are several different theories that explain this ionization process. One theory is that, as the liquid leaves the nozzle, the electric field induces a net charge on the small droplets. As the solvent evaporates, the droplet shrinks and the charge density at the surface of the droplet increases. The droplet finally reaches a point where the coulombic repulsion from this electric charge is greater than the surface tension holding it together. This causes the droplet to explode and produce multiply charged analyte ions. A typical ESI spectrum shows the distribution of molecular ions with different charge numbers.

Because electrospray produces multiply charged ions, high-molecular-weight compounds are observed at lower m/z value. This increases the mass range of the analyser so that high-molecular-weight compounds may be analysed with a less expensive mass spectrometer. An ion with a mass of 5000 and a charge of +10 is observed at *m/z* 500 and is easily analysed with an inexpensive quadrupole analyser.

Matrix-assisted laser desorption/ionization   Matrix-assisted laser desorption/ionization (MALDI) is used to analyse extremely large molecules. This technique directly ionizes and vaporizes the analyte from the condensed phase. MALDI is often used for the analysis of synthetic and natural polymers, proteins and peptides. The analysis of compounds with molecular weights up to 2,00,000 daltons is possible, and this high mass limit is continually increasing. In MALDI, both desorption and ionization are induced by a single laser pulse (Figure 6.7).

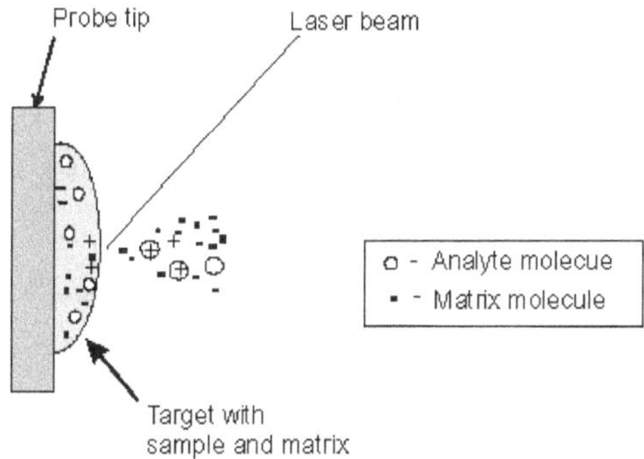

Figure 6.7   MALDI ionization

The sample is prepared by mixing the analyte and a matrix compound chosen to absorb the laser wavelength. This is placed on a probe tip and dried. A vacuum lock is used to insert the probe into the source region of the mass spectrometer. A laser beam is then focussed on this dried mixture and the energy from a laser pulse is absorbed by the matrix. This energy ejects analyte ions from the surface so that a mass spectrum is acquired for each laser pulse. The mechanism for this process is not well understood and is a subject of much controversy in the literature.

This technique is more universal (works with more compounds) than other laser ionization techniques in that the matrix absorbs the laser pulse. In other laser ionization techniques, the analyte must absorb at the laser wavelength. Typical MALDI spectra include the molecular ions, some multiply charged ions, and very few fragments.

An array of ionization methods and mass analysers are available to meet the needs of many types of chemical analysis. A few are listed in Table 6.2.

Table 6.2   Sample introduction/ionization method

Ionization method	Typical analytes	Sample introduction	Mass range	Features (or) Significance
Electron impact inonization (EI)	Relatively small, volatile	GC or liquid/solid probe	Up to 1000 daltons	Hard method, versatile, provides structure information.
Chemical ionization (CI)	Relatively small, volatile	GC or liquid/solid probe	Up to 1000 daltons	Soft method, molecular ion peak [MH]$^+$.
Electrospray ionization (ESI)	Peptides, proteins, non-volatile	Liquid chromatography or syringe	Up to 2,00,000 daltons	Soft method, ions often multiply charged.
Fast atom bombardment (FAB)	Carbohydrates, organometallics, peptides, non-volatile	Sample mixed in viscous matrix	Up to 6000 daltons	Soft method but harder than ESI or MALDI.
Matrix assisted laser desorption (MALDI)	Peptides, proteins, nucleotides	Sample mixed in solid matrix	Up to 5,00,000 daltons	Soft method, very high mass.

Ion Analyser

Molecular ions and fragment ions are accelerated by the manipulation of charged particles through the mass spectrometer. Uncharged molecules and fragments are pumped away. The quadrupole mass analyser uses positive (+) and negative (−) voltages to control the path of the ions. Ions travel down the path based on their mass-to-charge ratio (m/z). EI produces singly charged particles; hence the charge (z) is one. Therefore, the path of an ion will depend on its mass. If the (+) and (−) rods in the mass spectrometer were 'fixed' at a particular RF/DC voltage ratio, then one particular m/z would travel to the detector as shown by the solid line (the successful path) in Figure 6.8. However, voltages are not fixed but are scanned so that the ever-increasing masses can find a successful path through the rods to the detector.

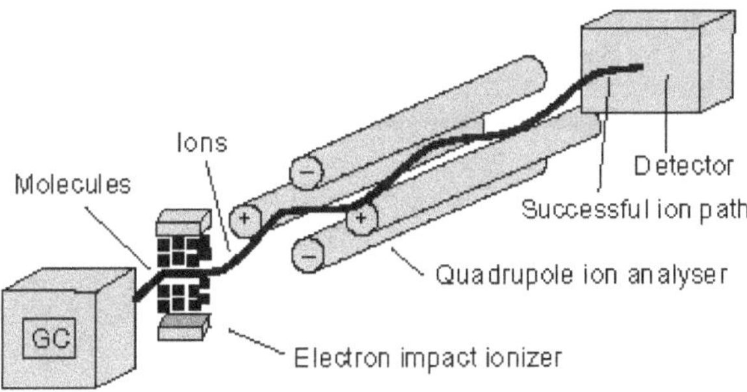

Figure 6.8   Quadrupole filter

The RF voltage rejects or transmits ions according to their m/z value by alternately focusing them in different planes (Figure 6.8). The four electrodes are connected in pairs and the RF potential is applied between these two pairs of electrodes. During the first part of the RF cycle, the top and bottom rods are at a positive potential; the left and right rods are at a negative potential. This squeezes positive ions into the horizontal plane. During the second half of the RF cycle, the polarity of the rods is reversed. This changes the electric field and focusses the ions in the vertical plane. The quadrupole field continues to alternate as the ions travel through the mass analyser. This causes the ions to undergo a complex set of motions that produces a three-dimensional wave.

The quadrupole field transmits selective ions because the amplitude of this three-dimensional wave depends upon the m/z value of the ion, the potentials applied, and the RF. By selecting an appropriate RF and RF potential, the quadrupole acts like a high pass filter, transmitting high m/z ions and rejecting low m/z ions. The low m/z ions have a greater acceleration rate; hence the wave for these ions has greater amplitude. If this amplitude is great enough, the ions will collide with the electrodes and cannot reach the detector. The low m/z value cut-off of the quadrupole is changed by adjusting the RF potential or the RF. Ions with a m/z greater than this cut-off are transmitted by the quadrupole.

Magnetic sector   The first mass spectrometer, built by J.J. Thompson in 1897, used a magnet to measure the m/z value of an electron. Magnetic sector instruments have evolved from this same concept. Sector instruments have higher resolution and greater mass range than quadrupole instruments, but they require larger vacuum pumps and often scan more slowly. The typical mass range is to m/z 5000, but this may be extended to m/z 30,000.

Magnetic sector instruments are often used in series with an electric sector (Figure 6.9) for high resolution and tandem mass spectrometry experiments.

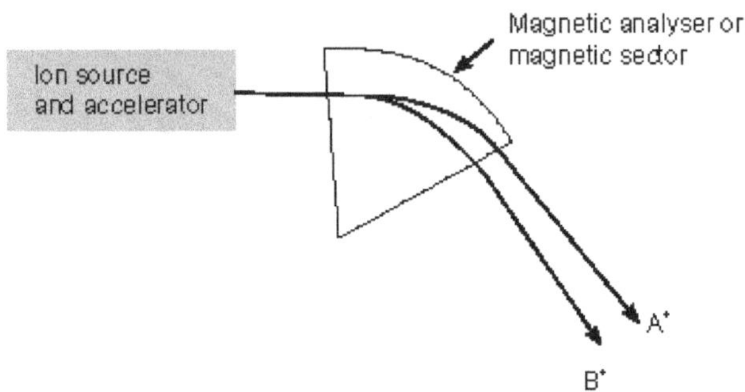

Figure 6.9   Magnetic sector mass spectrometer. The low *m/z* ion (B⁺) is separated from the high *m/z* ion (A⁺).

Magnetic sector instruments separate ions in a magnetic field according to the momentum and charge of the ion. Ions are accelerated from the source region into the magnetic sector by a 1 to 10 kV electric field. This acceleration is significantly greater than the 100 V acceleration typical for a quadrupole instrument. Since the ions are charged, as they move through the magnetic sector, the magnetic field bends the ion beam in an arc. This is the same principle that causes electric motors to turn. The radius of this arc (r) depends upon the momentum of the ion (μ), the charge of the ion (C) and the magnetic field strength (B).

$$r = \frac{\mu}{C \times B}$$

$$(6.1)$$

Ions with greater momentum will follow an arc with a larger radius. This separates ions according to their momentum; hence magnetic sectors are often called momentum analysers. The momentum of the ion is the product of the mass (m) and the velocity (v). The charge of the ion is the product of the charge number of the ion (z) and the charge of an electron (e). Substituting these variables into equation 1 yields:

$$r = \frac{m/z \times v}{B \times e}$$

$$(6.2)$$

The velocity of an ion is determined by the acceleration voltage (V) in the source region and the mass-to-charge ratio (m/z) of the ion. Equation 6.2 rearranges to give the m/z ion transmitted for a given radius, magnetic field and acceleration voltage as:

$$m/z = \frac{r^2 B^2 e}{2V}$$

$$(6.3)$$

Only one m/z value will satisfy equation 6.3 for a given radius, magnetic field and acceleration voltage. Other m/z ions will travel a different radius in the magnetic sector.

Older magnetic sector instruments used a photographic plate to simultaneously detect ions at different radii. Since each m/z has a different radius, they strike the photographic plate at a different location. Modern instruments have a set of slits at a fixed radius to transmit a single m/z to the detector. The mass spectrum is scanned by changing the magnetic field or the acceleration voltage to transmit different m/z ions. Some new instruments use multichannel diode array detectors to simultaneously detect ions over a range of m/z values.

Electric sector/double focusing mass spectrometers   An electric sector consists of two concentric curved plates. A voltage is applied across these plates to bend the ion beam as it travels through the analyser. The voltage is set so that the beam follows the curve of the analyser. The radius of the ion trajectory (r) depends upon the kinetic energy (V) of the ion and the potential field (E) applied across theplates.

$$r = \frac{2V}{E} \qquad (6.4)$$

Equation 6.4 shows that an electric sector will not separate ions accelerated to a uniform kinetic energy. The radius of the ion beam is independent of the ion's mass-to-charge ratio; hence the electric sector is not useful as a stand-alone mass analyser. An electric sector is, however, useful in series with a magnetic sector. The mass resolution of a magnetic sector is limited by the kinetic energy distribution (V) of the ion beam (equation 6.3). This kinetic energy distribution results from variations in the acceleration of ions produced at different locations in the source and from the initial kinetic energy distribution of the molecules. An electric sector significantly improves the resolution of the magnetic sector by reducing the kinetic energy distribution of the ions.

The effect of the electric sector is shown in Figure 6.10 for a reverse geometry (BE) instrument with the magnetic sector (B) located before the electric sector (E).

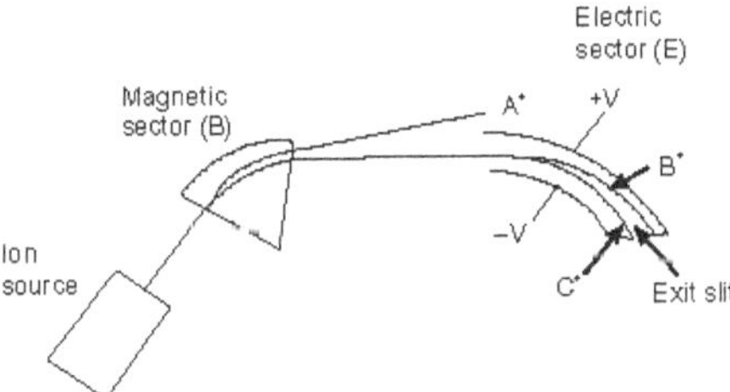

Figure 6.10   Reverse geometry double focusing mass spectrometer

A^+ is an m/z of 100.00 ion, B^+ is an m/z of 50.00 and C^+ is an m/z of 50.01. A^+ is rejected by the magnetic sector because of its greater momentum B^+ and C^+ are not resolved by the magnetic sector because they have the same momentum. To have the same momentum, however, B^+ must have a more kinetic energy than C^+. As a result the electric sector separates these two ions.

Time-of-flight mass analyser   The time-of-flight (TOF) mass analyser separates ions in time as they travel down a flight tube (Figure 6.11). This is a very simple mass spectrometer that uses fixed voltages and does not require a magnetic field. The greatest drawback is that TOF instruments have poor mass resolution, usually less than 500. These instruments have high transmission efficiency, no upper m/z limit, very low detection limits, and fast scan rates. For some applications, these advantages outweigh the low resolution. Recent developments in pulsed ionization techniques and new instrument designs with improved resolution have renewed interest in TOF-MS.

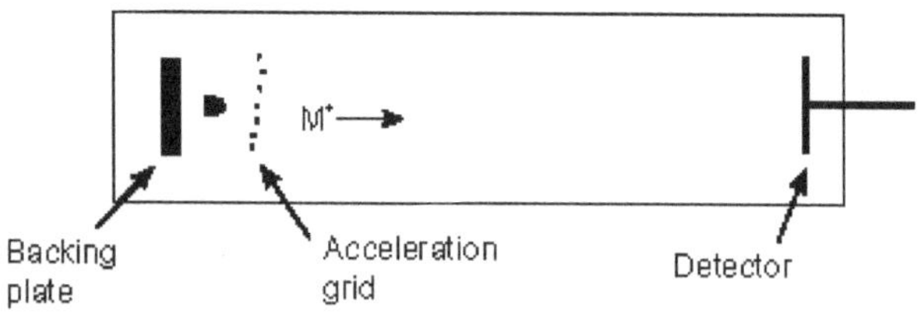

Figure 6.11   Time-of-flight mass spectrometer

In the source of a TOF analyser, a packet of ions is formed by a very fast (ns) ionization pulse. These ions are accelerated into the flight tube by an electric field (typically 2–25 kV) applied between the backing plate and the acceleration grid. Since all the ions are accelerated across the same distance by the same force, they have the same kinetic energy. Because velocity (v) is dependent upon the kinetic energy (E) and mass (m), lighter ions will travel faster.

$$E_{\text{kinetic}} = \frac{1}{2}mv^2$$

(6.5)

E_{kinetic} is determined by the acceleration voltage of the instrument (V) and the charge of the ion ($e \times z$). Equation 5 rearranges to give the velocity of an ion (v) as a function of acceleration voltage and m/z value.

$$v = \sqrt{\frac{2V \times e}{m/z}}$$

(6.6)

After the ions accelerate, they enter a 1- to 2-metre flight tube. The ions drift through this field-free region at the velocity reached during acceleration. At the end of the flight tube, they strike a detector. The time delay (t) from the formation of the ions to the time they reach the detector depends upon the length of the drift region (L), the mass-to-charge ratio of the ion, and the acceleration voltage in the source.

$$t = \frac{L}{2v \times e}\sqrt{m/z}$$

(6.7)

Equation 6.7 shows that low m/z ions will reach the detector first. The mass spectrum is obtained by measuring the detector signal as a function of time for each pulse of ions produced in the source region. Because all the ions are detected, TOF instruments have very high transmission efficiency, which increases the S/N level.

An example of a TOF mass spectrum is shown in Figure 6.12, where the mass of C60 is determined a 720 amu.

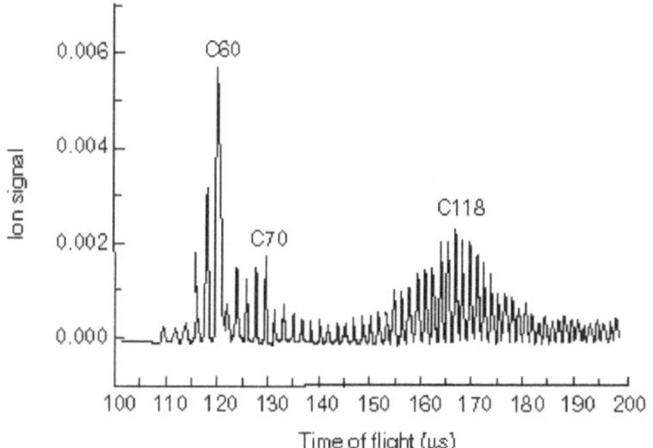

Figure 6.12 Laser ionization time-of-flight mass spectrum of C60 fullerene

Ion Detector

There are many types of detectors, but most work by producing an electronic signal when struck by an ion. Timing mechanisms, which integrate those signals with the scanning voltages, allow the instrument to report which m/z strikes the detector. The mass analyser sorts the ions according to m/z, and the detector records the abundance of each m/z. Regular calibration of the m/z scale is necessary to maintain accuracy in the instrument. Calibration is performed by introducing a well-known compound into the instrument and 'tweaking' the circuits so that the molecular ion and fragment ions of the compound are reported accurately.

Electron multiplier Perhaps the most common means of detecting ions involves an electron multiplier (Figure 6.13), which is made up of a series (12 to 24) of aluminum oxide (Al_2O_3) dynodes maintained at ever-increasing potentials. Ions strike the first dynode surface causing an emission of electrons. These electrons are then attracted to the next dynode held at a higher potential, and therefore more secondary electrons are generated. Ultimately, as numerous dynodes are involved, a cascade of electrons is formed, which results in an overall current gain in the order of one million or higher.

The high-energy dynode (HED) uses an accelerating electrostatic field to increase the velocity of ions. Since the signal on an electron multiplier is highly dependent on ion velocity, the HED serves to increase signal intensity and sensitivity.

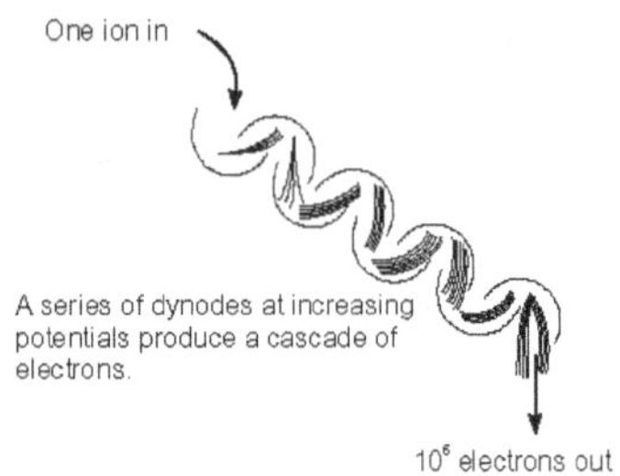

Figure 6.13 Diagrammatic representation of an electron multiplier and the cascade of electrons that results in a 10^6 amplification of current in a mass spectrometer

Faraday cup A Faraday cup (Figure 6.14) involves an ion striking the dynode (BeO, GaP or CsSb) surface, which causes secondary electrons to be ejected. This temporary electron emission induces a positive charge on the detector and, therefore, a current of electrons flowing towards the detector. This detector is not particularly sensitive, offering limited amplification of signal, yet it is tolerant of relatively high pressure.

Photomultiplier conversion dynode The photomultiplier conversion dynode detector (Figure 6.15) is not as commonly used as the electron multiplier, yet they are similar in design except that in the former the secondary electrons strike a phosphorus screen instead of a dynode. The phosphorus screen releases photons, which are detected by the photomultiplier. The photomultiplier operates much like an electron multiplier where the striking of the photon on a scintillating surface results in the release of electrons that are then amplified using the cascading principle. One advantage of the conversion dynod e is that the photomultiplier tube is sealed in a vacuum, unexposed to the environment of the mass spectrometer, and thus the possibility of contamination is less. This improves the life of these detectors over electron multipliers. A five-year lifetime is typical, and they have a sensitivity similar to that of an electron multiplier.

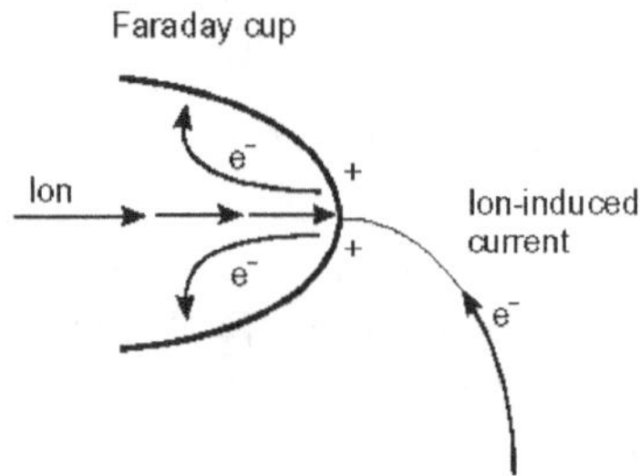

Figure 6.14 Faraday cup converts the striking ion into a current by temporarily emitting electrons creating a positive charge and the adsorption of the charge from the ion striking the detector

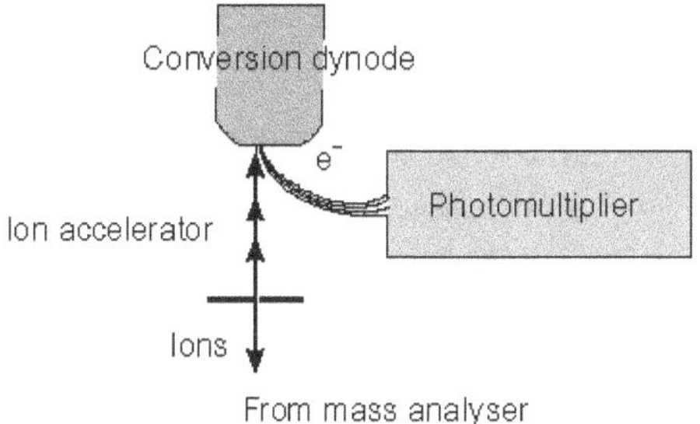

Figure 6.15 Photomultiplier with conversion dynode. Scintillation counting with a conversion dynode and a photomultiplier relies on the conversion of the ion (or electron) signal into light. Once the photon(s) are formed, detection is performed with a photomultiplier.

Array detector An array detector is a group of several detectors aligned in an array format. The array detector, which spatially detects ions according to their different m/z, has been typically used on magnetic sector mass analysers. Spatially differentiated ions can be detected simultaneously by an array detector. The primary advantage of an array detector is that, scanning is not necessary over a small mass range, and therefore sensitivity is improved.

Charge (or inductive) detector Charge detectors simply recognize a moving charged particle (an ion) through the induction of a current on the plate as the ion moves past. A typical signal is shown in Figure 6.16. This type of detection is widely used in FTMS to generate an image current of an ion. Detection is independent of ion size and, therefore, has been used on particles such as whole viruses.

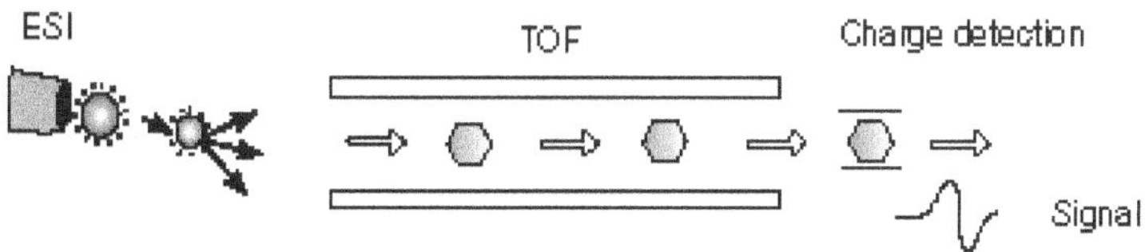

Figure 6.16 Illustration of the operation of a mass spectrometer with a charge detector; as the charged species passes through a plate, it induces a current on the plate.

The advantages and disadvantages of different types of detectors are listed in Table 6.3.

Table 6.3 Types of detectors, their advantages anddisadvantages

Detector	Advantages	Disadvantages
Electron multiplier	• Robust • Fast response • Sensitive ($\approx$ gain of 10^6)	• Shorter lifetime than scintillation counting ($\sim$3 years)
Photomultiplier conversion dynode (scintillation counting)	• Robust • Long lifetime (>5 years) • Sensitive ($\approx$ gains of 10^6)	• Cannot be exposed to light while in operation
Faraday cup	• Good for checking ion transmission and low sensitivity measurements	• Low amplification (≈ 10)
High-energy dynodes with electron multiplier	• Increases high mass sensitivity	• May shorten lifetime of electron multiplier
Array	• Fast and sensitive	• Reduces resolution • Expensive
Charge detector	• Detects ions independent of mass and velocity	• Limited compatibility with most existing instruments

General Mode of Fragmentation

The output of a mass spectrometer shows a plot of relative intensity vs. the mass-to-charge ratio (m/z). Mass-to-charge ratio is a dimensionless ratio of the mass number (m) of a given particle to the number (z) of the electrostatic unit (e) carried by the particle.

Some commonly used terms in mass spectrometry are given below:

Molecular ion   The ion obtained by the loss of an electron from the molecule. The symbol $\mathbf{M^+}$ is often used to represent molecular ion. The energy required to remove an electron varies in the order: lone pair < conjugated π < non-conjugated π < σ.

Base peak   The most intense peak in the mass spectrum, assigned 100% intensity. The intensities of other peaks including the parent peak are expressed as a percentage of base peak.

Radical cation   It is a positive-charged species with an odd number of electrons.

Metastable ions   The mass spectrum of a molecule generally shows sharp peaks at integral m/z values. Occasionally, diffuse, broad, low intensity and non-integral m/z

peaks can appear in a mass spectrum. These are due to metastable ions, which possess intermediate energy and have a sufficiently long lifetime to leave the ion source, but they may decompose during flight before detection. They have been accelerated as one species (metastable ion) and resolved as another (the daughter ion).

$$A^+ \longrightarrow B^+ \; + \; C$$

$$\text{Accelerated} \qquad \text{Detected} \quad \text{Neutral}$$

The precursor ion A^+ is sometimes called **parent ion.**

Kinetic factors are also important in the fragmentation process. The rate constant for fragmentation increases with internal energy to a limiting rate of 10^{13} s^{-1}. Since ions spend approximately 10^{-6} s in the ion source prior to acceleration, only those reactions with rate constant in the range $10^6 - 10^{13}$ s^{-1} will be observed as normal ions in the mass spectrum. The ions with rate constants in the range 10^4 to 10^6 s^{-1} will dissociate and may be detected as weak, diffuse metastable ions.

Doubly charged ions   Peaks pertaining to stable doubly charged ions may be useful in the interpretation of mass spectrum. The molecular ion that becomes doubly charged during the ionization process will generate a peak at an m/z value numerically equal to one-half the molecular weight of the compound. For example, the mass spectrum of chrysene (MW 228) exhibits a relatively intense peak at m/z 114 for the doubly charged molecular ion.

Fragment ions   The lighter cations formed by the decomposition of molecular ion. These often correspond to stable carbocations (Table 6.4).

Table 6.4   Common mass spectrum fragments

Commonly lost fragments		Common stable ions
m-15	·CH$_3$	
m-17	·OH	m/z = 43 CH$_3$—C≡O
m-26	·CN	
m-28	H$_2$C=CH$_2$	
m-29	·CH$_2$CH$_3$ ·CHO	m/z = 91
m-31	·OCH$_3$	
m-35	Cl	
m-43	·CH$_3$Ċ=O	m/z = m-1
m-45	OCH$_2$CH$_3$	
m-91		

The fragmentation of molecular ions into an assortment of fragment ions is a mixed blessing. The nature of the fragment ions often provides a clue to the molecular structure, but if the molecular ion has a life of less than a few microseconds, it will not survive long enough to be observed. Without a molecular ion peak as a reference, the difficulty in interpreting a mass spectrum increases markedly. Fortunately, most organic compounds give mass spectra that include a molecular ion, and others, which do not give mass spectra, often respond successfully to milder ionization conditions. Among simple organic compounds, the most stable molecular ions are those from aromatic rings, other conjugated pi-electron systems and cycloalkanes. Alcohols, ethers and highly branched alkanes generally show the greatest tendency towards fragmentation.

The mass spectrum of dodecane (Figure 6.17) illustrates the behaviour of an unbranched alkane. Since there are no heteroatoms in this molecule, there are no non-bonding valence shell electrons. Consequently, the radical cation character of the molecular ion (m/z = 170) is delocalized over all the covalent bonds.

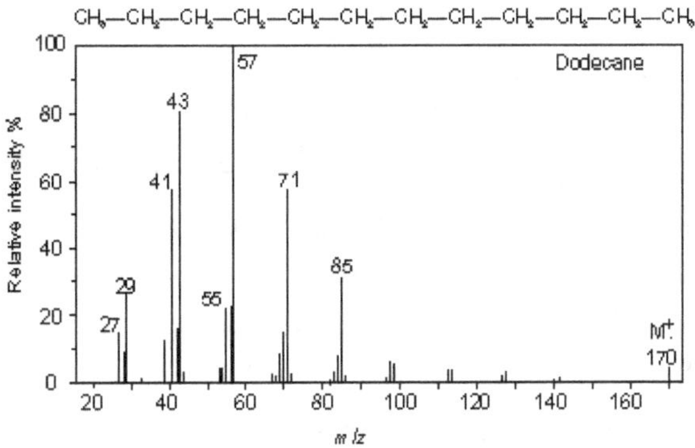

Figure 6.17   Mass spectrum of dodecane

Fragmentation of C—C bonds occurs because they are usually weaker than C—H bonds, and this produces a mixture of alkyl radicals and alkyl carbocations. The positive charge commonly resides on the smaller fragment, and so we see a homologous series of hexyl (m/z = 85), pentyl (m/z = 71), butyl (m/z = 57), propyl (m/z = 43), ethyl (m/z = 29) and methyl (m/z = 15) cations. These are accompanied by a set of corresponding alkenyl carbocations (e.g. m/z = 55, 41 and 27) formed by loss of 2 H. All the significant fragment ions in this spectrum are even-electron ions. In most alkane spectra, the propyl and butyl ions are in abundance.

The presence of a functional group, particularly one having a heteroatom Y with non-bonding valence electrons (Y = N, O, S, X, etc.), can dramatically alter the fragmentation pattern of a compound. This influence is thought to occur because of localization of the radical cation component of the molecular ion on the heteroatom. After all, it is easier to remove (ionize) a non-bonding electron than one that is part of a covalent bond. By localizing the reactive moiety, certain fragmentation processes will be favoured. These are summarized in Figure 6.18, where the shaded box at the top displays examples of such localized molecular ions. The first two fragmentation paths lead to even-electron ions, and the elimination (path number 3) gives an odd-electron ion. Note the use of different curved arrows to show single electron shifts as compared with eletron pair shifts.

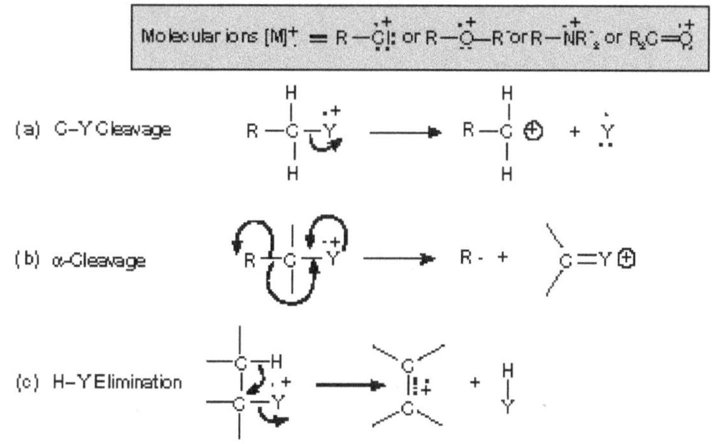

(a) C–Y Cleavage

(b) α-Cleavage

(c) H–Y Elimination

Figure 6.18 Fragmentation paths

The charge distributions shown in Figure 6.18 are common, but for each cleavage process the charge may sometimes be carried by the other (neutral) species, and both the fragment ions are observed. Of the three cleavage reactions described here, the alpha-cleavage is generally favoured for nitrogen, oxygen and sulphur compounds. Indeed, in the spectra of 4-methyl 3-pentene-2-one and N,N-diethylmethylamine (Figures 6.19 and 6.20) the major fragment ions come from alpha-cleavages.

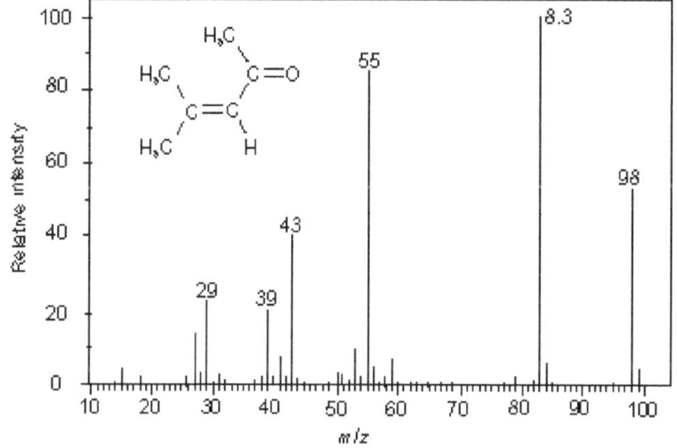

Figure 6.19 Mass spectrum of 4-methyl 3-pentene-2-one

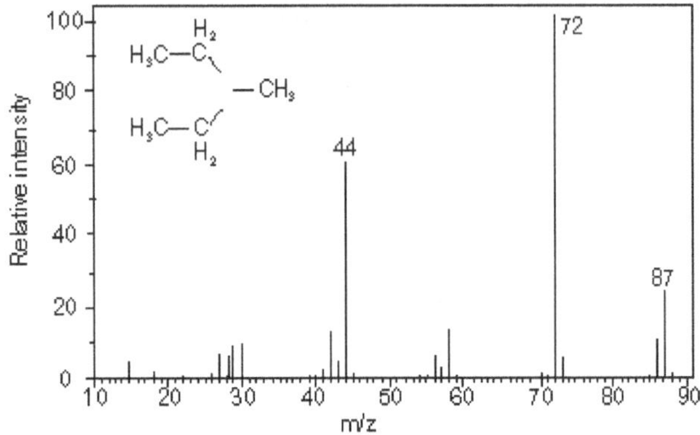

Figure 6.20   Mass spectrum of N,N-diethylmethylamine

Alkanes   Simple alkanes tend to undergo fragmentation (Figure 6.21) by the initial loss of a methyl group to form a (m-15) species. This carbocation can then undergo stepwise cleavage down the alkyl chain, expelling neutral two-carbon units (ethene). Branched hydrocarbons form more stable secondary and tertiary carbocations, and these peaks will tend to dominatethe mass spectrum.

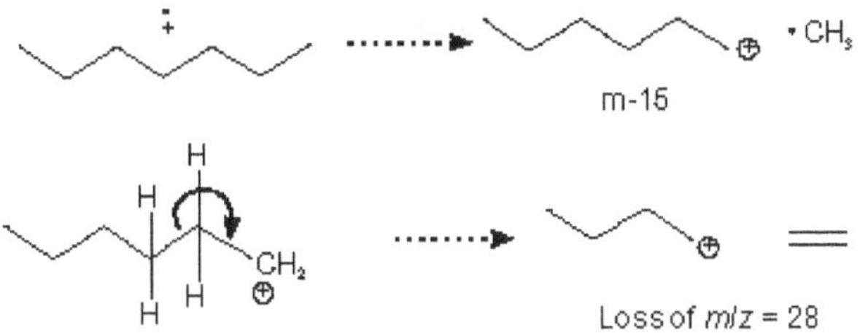

Figure 6.21   Fragmentation pattern in alkanes

Aromatic hydrocarbons   The fragmentation of aromatic nucleus (Figure 6.22) is somewhat complex, generating a series of peaks having m/z = 77, 65, 63, etc. While these peaks are difficult to describe in simple terms, they do form a pattern (the 'aromatic cluster') that becomes recognizable with experience. If the molecule contains a benzyl unit, the major cleavage will be to generate the benzyl carbocation, which rearranges to form the tropylium ion. Expulsion of acetylene (ethyne) from this generates a characteristic m/z = 65 peak.

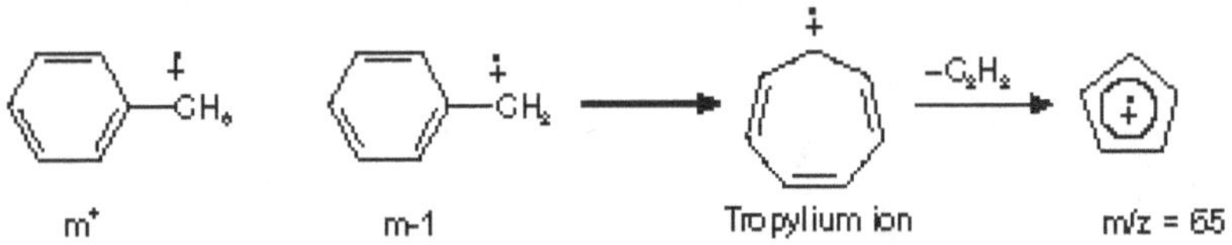

Figure 6.22   Fragmentation in aromatics with benzyl unit

Aldehydes and ketones   The predominant cleavage in aldehydes and ketones is loss of one of the side-chains to generate the substituted oxonium ion. This is an extremely favourable cleavage and this ion often represents the base peak in the spectrum. The methyl derivative $(CH_3C \equiv O^+)$ is commonly referred to as the

Another common fragmentation observed in carbonyl compounds (and in nitriles, etc.) involves the expulsion of neutral ethene via a process known as McLafferty rearrangement, following the general mechanism shown below.

Esters, acids and amides   As with aldehydes and ketones, the major cleavage observed for these compounds involves expulsion of the 'X' group, as shown below, to form the substituted oxonium ion. For carboxylic acids and unsubstituted amides, characteristic peaks at m/z = 45 and 44 are also often observed.

Alcohols   In addition to losing a proton and hydroxy radical, alcohols tend to lose one of the α-alkyl groups (or hydrogens) to form oxonium ions as shown below. For primary alcohols, this generates a peak at m/z = 31; secondary alcohols generate peaks with m/z = 45, 59, 73, etc., according to substitution.

Ethers   Following the trend of alcohols, ethers will fragment, often by loss of an alkyl radical, to form a substituted oxonium ion, as shown below for diethyl ether.

Halides   Organic halides fragment with simple expulsion of the halogen, as shown below. The molecular ions of chlorine- and bromine-containing compounds will show multiple peaks due to the fact that each of these exists as two isotopes in relatively high abundance. Thus, for chlorine, the $^{35}Cl/^{37}Cl$ ratio is roughly 3.08:1, and for bromine, the $^{79}Br/^{81}Br$ ratio is 1.02:1. The molecular ion of a chlorine-containing compound will have two peaks, separated by two mass units, in the ratio $\approx$ 3:1 and of a bromine-containing compound will have two peaks, again separated by two mass units having approximately equal intensities

Nitrogen rule   An organic compound containing an odd number of nitrogen atoms will have a molecular ion with an odd mass number, provided the other elements present are carbon, hydrogen, oxygen, sulphur, silicon or halogens. Similarly, a molecule with an even-molecular-weight contains no nitrogen or an even number of nitrogen atoms.

The nitrogen rule can be used because nitrogen has an even atomic weight and an odd valence whereas all other elements encountered in organic mass spectrometry have either an even mass and even valence or an odd mass and odd valence. The inclusion of any common stable isotope, except ^{18}O, alters the use of nitrogen rule.

Even-electron rule   This rule predicts the most likely sequence of ions. It states that an even-electron species will not normally fragment into two odd electron species since the total energy of this product mixture would be too high. In preference, an ion will degrade to another ion and a neutral molecule.

$$A^+ \longrightarrow C^+ + N_1$$

$$A^+ \text{---} X \longrightarrow D^{+\bullet} + N_2$$

Radical ions, being odd-electron species, undergo fragmentation by loss of radicals or even-electron molecules.

$$B^{+\bullet} \longrightarrow E^+ + N_3$$

$$B^{+\bullet} \longrightarrow F^+ + N_4$$

where

B^+ is a molecular ion or daughter ion formed by the loss of a neutral molecule from the molecule ion,

$N_2^{\bullet}$ and $N_3^{\bullet}$ are neutral radicals and

N_1 and N_4 are neutral molecules.

Chromatography–Mass Spectrometry

Gas Chromatography–Mass Spectrometry (GC–MS)

GC–MS involves compatible analytical techniques that are well matched in almost every aspect of gas chromatography except that a gas chromatograph functions with a column outlet maintained at one atmosphere whereas a mass spectrometer operates at source pressures between 10^{-5} and 10^{-6} torr. GC–MS works for neutral molecules containing relative molecular mass up to about 1500. Direct coupling of a GC with MS can be usually achieved through an interface (separator). These devices increase sample concentration and allow the GC effluent to be introduced into high vacuum prevalent in the ion source of a mass spectrometer.

The problem with the high pressure of the carrier gas in the column effluent and with the high dilution of the sample entering the mass spectrometer can be solved with enrichers. These include fritted tube separator, all-glass jet system and the permeable-membrane type separator.

The permeable-membrane-type separator utilizes a membrane that selectively passes soluble organic molecules and acts as a barrier to helium. Modern mass spectrometers are capable of determining nanogram quantities and even up to picogram level.

Applications   The GC–MS is used

1.   in flavour chemistry where the aroma of a substance is found to consist of a mixture of several hundred relatively simple compounds.

2.   to separate metabolites from many natural compounds in order to study drug metabolism.

3.   to analyse pesticides and pollutants.

Liquid Chromatography–Mass Spectrometry (LC–MS)

A mass spectrometer has the potential to serve as a universal detector for high performance liquid chromatography (HPLC). The following difficulties may arise when LC and MS are combined.

1.   It is hard to deal with liquid solvent. Even though the flow rates are small for microcolumns, they will represent a major problem in maintaining an operational vacuum in mass spectrometer.

2.   LC is usually reserved for those polar compounds that are not amenable to GC and, in many cases, to conventional mass spectrometry with EI or CI.

Electrospray Ionization (ESI)

Description   The HPLC line is connected to the electrospray probe which consists of a metallic capillary surrounded by nitrogen flow. A voltage is applied between the probe tip and the sampling cone. In most instruments, the voltage is applied on the capillary, while the sampling cone is held at low voltage. The first step is to create a spray. At very low flow rate (a few μL/min), the difference in potential is sufficient to create the spray. At higher flow rate, nitrogen flow is necessary to maintain a stable spray.

The API sources include a heating device to speed up solvent evaporation.

A mandatory condition to work with electrospray is that the compound of interest must be ionized in a solution.

If it is not compatible with the HPLC conditions, i.e., in the case of normal phase chromatography, it is possible to use post column addition to get appropriate conditions.

Due to the electric field at the tip of the capillary, the surface of the droplets containing ionized compound will get charged, either positively or negatively, depending on voltage polarity. Because of solvent evaporation, the size of the droplet reduces, and consequently the density of charges at the droplet surface increases. The repulsion forces between the charges increase until the droplet explodes. This process repeats until analyte ions evaporate from the droplet.

Multiply charged ions can be obtained depending on the chemical structure of the analyte and so ESI is the technique of choice for analysing proteins and other biopolymers on quadrupole or ion trap analysers (Figure 6.23).

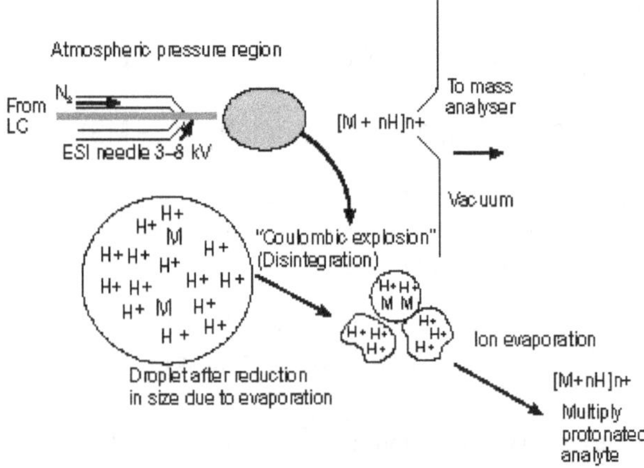

Figure 6.23   Electrospray ionization

Typical ions produced by electrospray ionization	
Positive mode:	[M+H]+ protonated molecule [M+Na]+, [M+K]+... adducts [M+CH$_3$CN+H] + protonated, + solvent adducts
Negative mode:	[M-H]- deprotonated molecule [M+HCOO-]-,... adducts

Operating conditions

Flow rate   The best sensitivity is achieved at low flow rate. Working at 1 mL/min or even higher is technically possible, but may cause a reduction in the signal-to-noise ratio.

Eluent pH   The mobile phase should have a pH such that the analytes are ionized. An acidic mobile phase is suitable for the analysis of basic compounds, using positive ESI, while a basic pH will be chosen for analysing acidic molecules. However, some exceptions exist to this general rule: positive ESI of basic compounds with a high pH mobile phase has been published.

Buffers   Volatile buffers are preferred for routine use. Operating the instrument with non-volatile buffers, such as phosphate, is technically possible, but the salt deposit in the source will have to be removed periodically. The concentration of the buffer, or the acid or base used to adjust/control the pH, should be as low as possible. If not, competition between analyte and electrolyte ions for conversion to gas phase ions decreases the analyte response. Let us take this point further: if a species is in large excess, it will cover the droplet surface and will prevent other ions from accessing the surface and will obstruct evaporation. A species in large excess will also catch all charges available and also prevent the ionization of other molecules present at much lower concentration.

Ion pairing agents   These molecules (e.g. sodium octanesulphonates) have surfactant properties. The presence of surfactants in the mobile phase impacts spray formation and droplet evaporation. There is also a surface competition mechanism.

Matrix effects   When the sample contains high concentration of salts, or an excess of another analyte that can ionize in the operating condition, there might be a competition effect in ionization. This is called **ion suppression**. The chromatographic separation must be developed to remove this effect, at least when doing quantitative analysis The block diagram of LC–MS is shown in Figure 6.24.

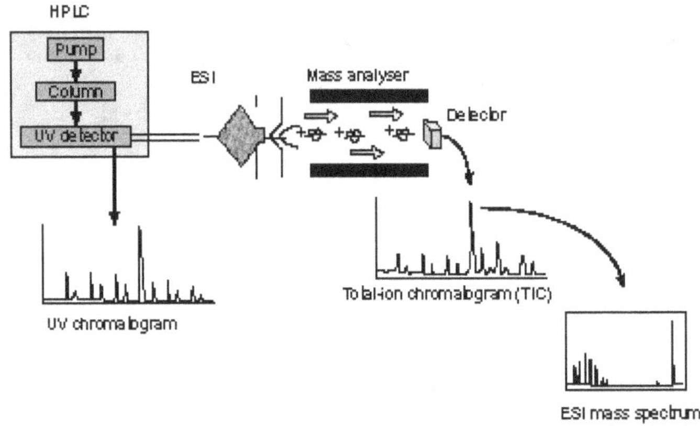

Figure 6.24   Liquid chromatography–mass spectrometry

Negative Ion Mass Spectrometry

Molecular negative ions are produced by the low-energy electron capture process, $AB + e^- \rightarrow AB^-$, and fragment negative ions are formed by the dissociative process, $AB \rightarrow [AB^-]^* \rightarrow A^- + B$. Negative molecular ions are formed by the capture of low-energy secondary electrons, originating at either metal surfaces or from gas-phase positive ionization (i.e., $AB + e^- \rightarrow AB^+ + 2e^-$). Primary electron energies of 70 eV are used for this purpose.

Negative chemical ionization (NCI) typically requires an analyte (Figure 6.25) that contains electron-capturing moieties (e.g. fluorine atoms or nitrobenzyl groups). Such moieties significantly increase the sensitivity of NCI, in some cases, 100 to 1000 times greater than that of electron ionization (EI). NCI is probably one of the most sensitive techniques and is used for a wide variety of small molecules with the caveat that the molecules are often chemically modified with an electron-capturing moiety prior to analysis.

Low-energy electrons (2 sto 4 eV) from an argon discharge tube can increase the formation of negative ions upon impact. From the spectra of a large number of compounds, it has been observed that ions with masses near the molecular ion such as M^- (resonance capture), M-1 (loss of hydrogen), and M+16 and M-1)+16 predominated.

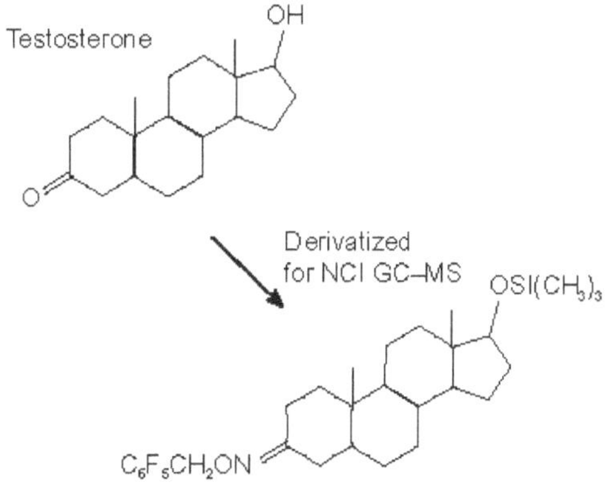

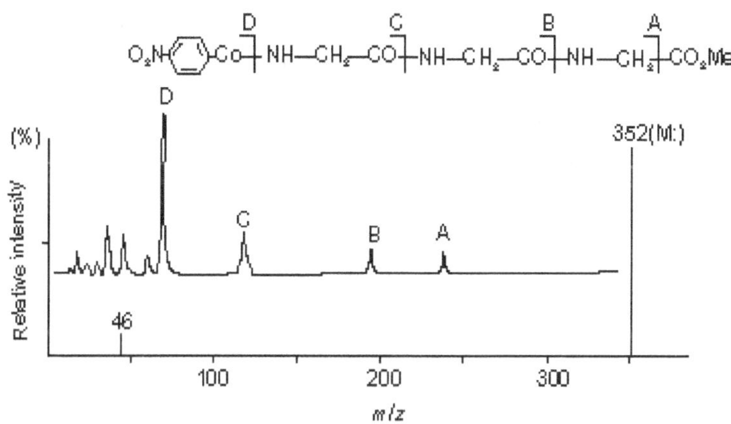

Figure 6.25　Chemical derivatization to enhance vaporization and ionization. The pentafluorobenzyl trimethyl silyl ether derivatives of steroids make them more amenable to high-sensitivity measurements using negative chemical ionization.

Finally, the charge inversion spectrum of a negative ion may also be used for the determination of structure of the unknown molecules. A molecule may sometimes form a molecular anion by secondary-electron capture), or an M-1 ion by OH⁻, which does not fragment. Alternatively, a compound that does not yield negative ions may form a stable molecular negative ion when an electron capture group (e.g. nitrophenyl or perfluorophenyl) is specifically built into that molecule. The structural information from fragmentation data is not available in any of these cases. The measurement of the collision-induced charge inversion spectrum of the molecular negative ion (or (M-1)⁻ ion when appropriate) gives the required sructural information.

Figure 6.26　Negative ion mass spectrum of dipeptide derivative

Experimentally, all that needs to be done to measure the negative ion spectrum to give the molecular weight, introduce nitrogen into the collision region, and reverse the polarity of the magnet to allow the measurement of positive ions. This technique, which constitutes an effective marriage of negative and positive ion mass spectrometry, has been applied to amino acids and peptides, and to the NCI spectra of mixtures of organic compounds. An example of this method is shown for a dipeptide derivative in Figure 6.26. The negative ion spectrum is dominated by the molecular negative ion; the charge inversion spectrum determines the structural features.

Applications of Mass Spectrometry

Structure Determination

Mass spectrometry can be used to determine the formulae and structures of organic molecules. The process of identification proceeds in the following order.

1. *Identification of molecular ion*   The mass of the molecular ion is calculated from the molecular formula by using the atomic weights of the isotopes with the highest natural abundance.

2. *Study of the isotope distribution pattern*   This is the pattern of peaks due to the combinations of different atomic isotopes in a specific molecule or fragment.

3. *Explaining the fragmentation pattern*   When the sample molecules are ionized in the source, some of the energy introduced to cause ionization serves to break up the molecule. The resulting fragments help in confirming the identification of the molecule.

Determination of Molecular Weight and Molecular Formula

A low-resolution mass spectrometer measures m/z values to the nearest whole number of mass unit. However, a high-resolution mass spectrometer measures m/z values to three or four decimal places, and thus provides an accurate method for determining the molecular weights.

High-resolution mass spectrometry

Formula	C_6H_{12}	C_5H_8O	$C_4H_8N_2$
Mass	84.0939	84.0575	84.0688

In assigning mass values to atoms and molecules, we have assumed integral values for isotopic masses. However, accurate measurements show that this is not strictly true. Because the strong nuclear forces that bind the components of an atomic nucleus together vary, the actual mass of a given isotope deviates from its nominal integer by a small but characteristic amount ($E = mc^2$). Thus, relative to ^{12}C at 12.0000, the isotopic mass of ^{16}O is 15.9949 amu (not 16) and ^{14}N is 14.0031 amu (not 14). By designing mass spectrometers that can determine m/z values accurately to four decimal places, it is possible to distinguish different formulae having the same nominal mass. The above table illustrates this important feature, and a double-focusing high-resolution mass spectrometer can easily distinguish ions having these compositions. Mass spectrometry, therefore, not only provides a specific molecular mass value but also establishes the molecular formula of an unknown compound.

A carbonyl complex of iron gives a molecular ion at 504 amu. Because iron has a mass of 56, that total mass can be fitted equally well by the formulae $Fe(CO)_{16}$, $Fe_2(CO)_{14}$, $Fe_3(CO)_{12}$, $Fe_4(CO)_{10}$, $Fe_5(CO)_8$, and so on. However, with high-resolution instruments, it is possible to determine the masses of ions to within a few parts per million, and as atomic masses are not exact integers, it is usually possible to distinguish between the various options. The exact masses for $Fe_3(CO)_{12}$ and $Fe_4(CO)_{10}$ are 503.7438 and 503.6889 amu, respectively. Hence, it is easy to decide which formula is correct.

Environmental and Medical Applications

Polychlorinated dibenzo-*p*-dioxins (PCDDs) and polychlorinated dibenzo-*p*-furans (PCDFs) exhibit toxic effects. There are 75 possible PCDD congeners, and the most dangerous is 2,3,7,8-tetrachlorobenzo-*p*-dioxin (TCDD). Sensitivity and selectivity are the two most important factors in utilizing mass spectrometry in such studies. Preconcentration of sample is often necessary prior to analysis. GC–MS is widely employed for the analysis of a large number of drug compounds and their metabolites.

Isotope Patterns

Mass spectrometers are capable of separating and detecting individual ions, even those that only differ by a single atomic mass unit. As a result, molecules containing different isotopes can be distinguished. This is most apparent when atoms such as bromine or chlorine are present (^{79}Br:^{81}Br, intensity 1:1; ^{35}Cl:^{37}Cl, intensity 3:1) where peaks at **M** and **M+2** are obtained. The intensity ratios in the isotope patterns are due to the natural abundance of isotopes. **M+1** peaks are seen due the presence of ^{13}C in the sample.

The following two mass spectra are examples of haloalkanes with characteristic isotope patterns.

The first mass spectrum (Figure 6.27) is of 2-chloropropane. Note the isotope pattern at 78 and 80, which represent the **M** and **M+2** in a 3:1 ratio. Loss of ^{35}Cl from 78 or ^{37}Cl from 80 gives the base peak at $m/z = 43$, corresponding to the secondary propyl cation. Note that the peaks at m/z = 63 and 65 still contain Cl and, therefore, also show the 3:1 isotope pattern.

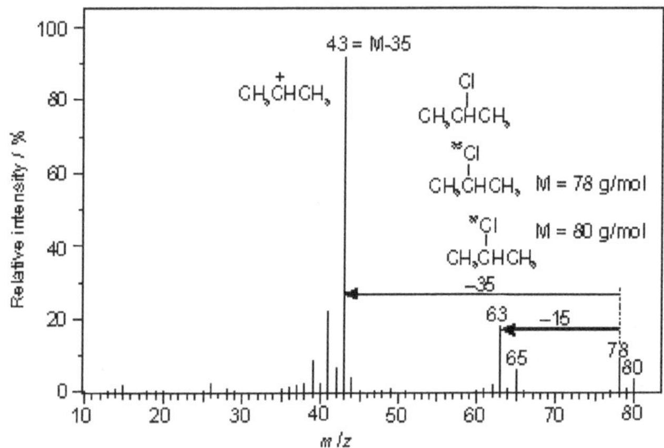

Figure 6.27   Mass spectrum of 2-chloropropane

The second mass spectrum (Figure 6.28) is of 1-bromopropane. Note the isotope pattern at 122 and 124 that represent the M and M+2 in a 1:1 ratio. Loss of ^{79}Br from 122 or ^{81}Br from 124 gives the base peak at m/z = 43, corresponding to the propyl cation. Note that the other peaks, such as those at m/z = 107 and 109, still contain Br and, therefore, also show the 1:1 isotope pattern.

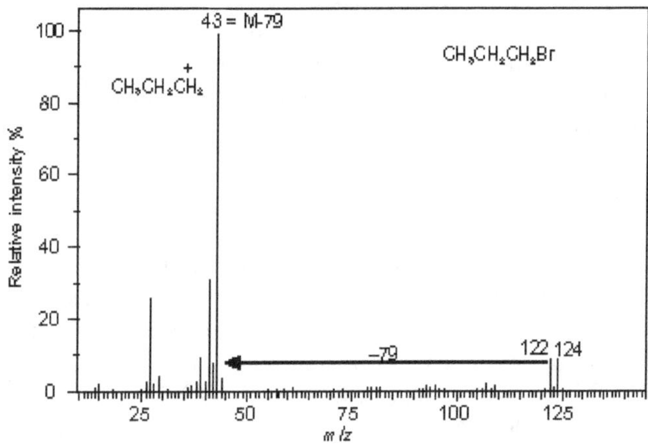

Figure 6.28   Mass spectrum of 2-bromopropane

The complexity of fragmentation patterns has led to mass spectra being used as fingerprints for identifying compounds. Environmental pollutants, pesticide residues on food and substance identification of are a few examples of this application. Extremely small amount of an unknown substance sample (a microgram or lesser) are sufficient for such analysis.

The following mass spectrum of cocaine (Figure 6.29) demonstrates how a forensic laboratory might determine the nature of an unknown street drug. Even though extensive fragmentation has occurred, most of the abundant ions (identified by magenta numbers) can be rationalized by the three mechanisms described earlier. Possible assignments may be seen by clicking on the spectrum, and it should be noted that all are even-electron ions. The m/z = 42 ion might be any or all of the following: C_3H_6, C_2H_2O or C_2H_4N. A precise assignment could be made from a high-resolution m/z value.

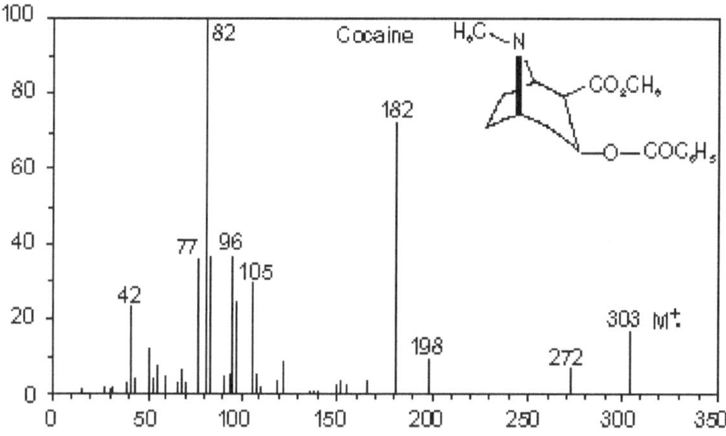

Figure 6.29   Mass spectrum of cocaine

Odd-electron fragment ions are often formed by characteristic rearrangements in which stable neutral fragments are lost. Mechanisms for some of these rearrangements have been identified by following the course of isotopically labelled molecular ions. A few examples of these rearrangement mechanisms may be seen by clicking the following button.

Problems and Solutions

Analyse $C_5H_{12}O$.

- ☀ MW: 8815
- ☀ Mass spectrum:
- ☀ Structure:

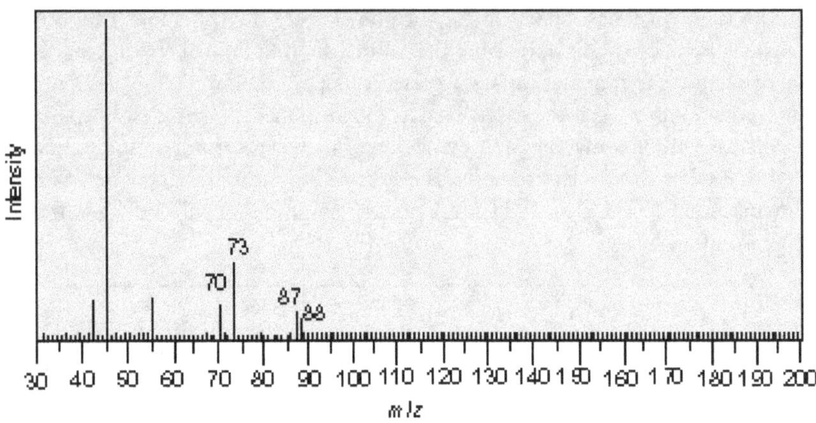

* IUPAC name: 2-pentanol
* MS fragments:

The spectrum shows a small molecular ion and a small peak resulting from loss of a hydrogen atom from alcohol. The loss of a methyl group gives the m-15 peak and loss of hydroxy radical gives the secondary carbocation at m-17. The base peak is formed by expulsion of the alkyl chain to give the simple oxonium ion at m/z = 45.

Hexane shows the same fragmentation pattern as other unbranched alkanes. Thus, alkyl carbocations at m/z = 15, 29, 43 and 57 amu provide dominant peaks in the spectrum. The m/z = 57 butyl cation (M-29) is the base peak, and the m/z = 43 and m/z = 29 ions are also abundant.

Chain branching clearly influences the fragmentation of this isomeric hexane. The molecular ion at m/z = 86 is weaker than that for hexane itself and the M-15 ion at m/z = 71 is stronger. The molecular ion at m/z = 57 is almost absent (try to find a simple cleavage that gives a butyl group). An isopropyl cation at m/z = 43 is very strong, and the corresponding propene radical-cation at m/z = 42 (coloured orange), produced by loss of propane, gives the base peak.

By having six carbons of hexane close to a ring, the fragmentation is profoundly changed. To begin with, the molecular ion at m/z = 84 (Figure 6.30) is much stronger than the corresponding ions in the previous acyclic compounds. The base peak at m/z = 56 is produced by loss of ethene, and so it is an odd-electron ion (coloured orange). The alkenyl cations at m/z = 41 and 27 are stronger than the corresponding alkyl cations at m/z = 43 and 29. The loss of methyl at m/z = 69 and a corresponding small molecular ion at m/z = 15 obviously requires some hydrogen rearrangements.

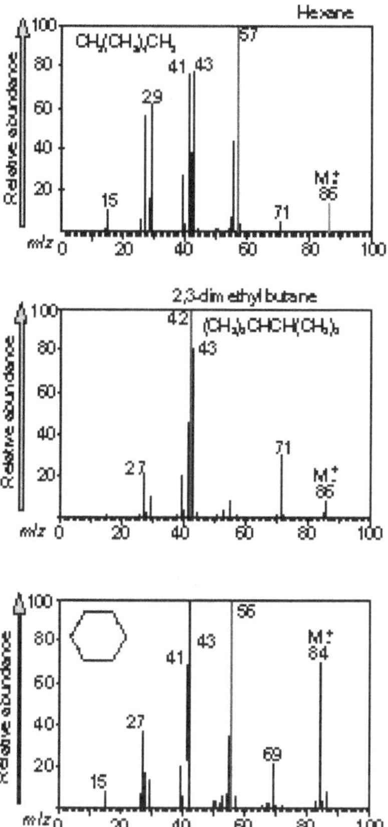

Figure 6.30   Hydrocarbons having no functional groups

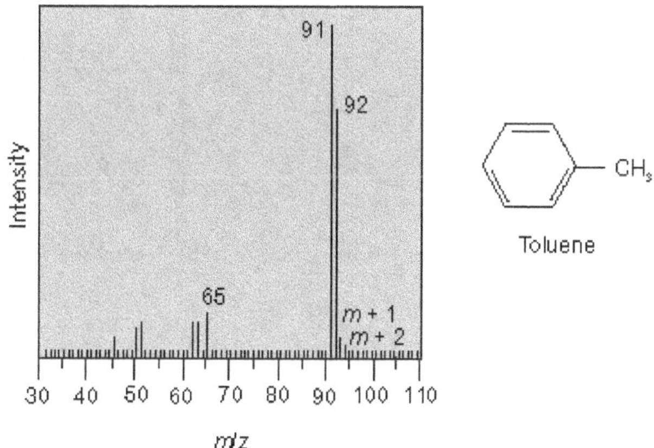

Figure 6.31   Mass spectrum of toluene

The mass spectrum of toluene (methyl benzene) is shown in Figure 6.31. The spectrum displays a strong molecular ion at m/z = 92, small (m+1) and (m+2) peaks, a base peak at m/z = 91 and an assortment of minor peak m/z = 65 and below.

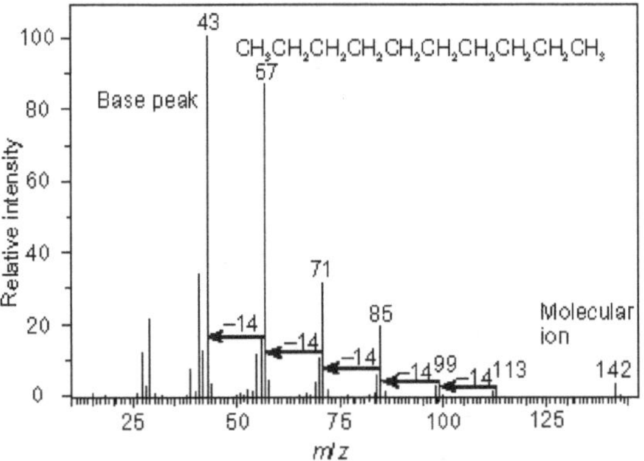

Figure 6.32   Mass spectrum of decane

The mass spectrum of a typical hydrocarbon, n-decane, is shown in Figure 6.32. The **molecular ion** is seen as a small peak at m/z = 142. Notice the series of ions detected that correspond to fragments, which differ by 14 mass units, formed by the cleavage of bonds at sucessive —CH_2— units.

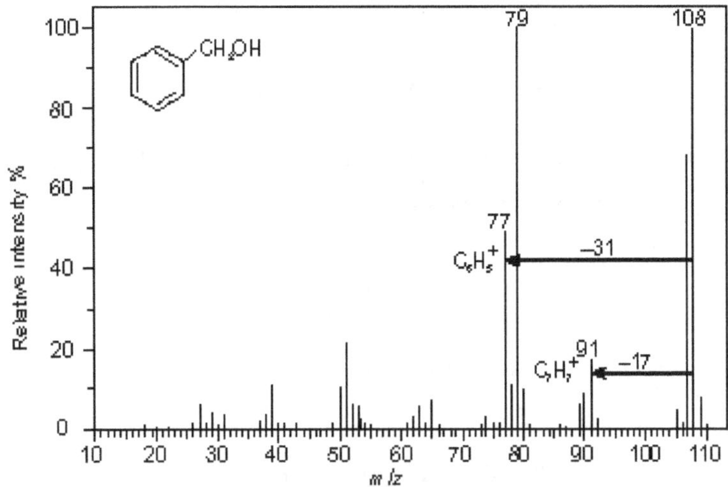

Figure 6.33   Mass spectrum of benzyl alcohol

The mass spectrum of benzyl alcohol is shown in Figure 6.33. The molecular ion is seen at m/z = 108. Fragmentation via loss of 17 (—OH) gives a common fragment as seen for alkyl benzenes at m/z = 91. Loss of

31 ($-CH_2OH$) from the molecular ion gives a fragment at $m/z = 77$ corresponding to the phenyl cation. The small peaks at 109 and 110, correspond to the presence of small amounts of ^{13}C in the sample (which has about 1 natural abundance).

REVIEW QUESTIONS

1. What is meant by base peak?

2. Explain nitrogen rule.

3. Discuss the importance of metastable peaks.

4. What is McLafferty rearrangement?

5. Discuss the basic principle of mass spectrometry.

6. Outline the important features of the mass spectrum of alcohols.

7. What is the use of molecular ion peak?

8. Describe the instrumentation of mass spectrometer.

9. Write in detail the general fragmentation modes in mass spectrometry.

10. Discuss the important features of mass spectra of alkanes and alkenes with suitable examples.

11. Why are isotope peaks present in the mass spectrum of a compound?

12. Discuss the applications of mass spectrometry.

13. Write notes on:

    i. Ion analysers

    ii. Detectors

    iii. Resolution

Chapter 7

ELECTRON SPIN RESONANCE SPECTROSCOPY

Introduction

Electron paramagnetic resonance (EPR) or electron spin resonance (ESR) is defined as the form of spectroscopy concerned with microwave-induced transitions between the magnetic energy levels of electrons having a net spin and orbital angular momentum. In the present context, the magnetic field scanning method is assumed. Other methods, however, are conceivable.

Electron paramagnetic resonance (EPR) is a sensitive spectroscopic method for the determination of the geometric and electronic structure, the dynamics and the spatial distribution of paramagnetic species in materials. EPR directly focuses on the unpaired electrons and nuclei in their vicinity and is therefore the method of choice for studying free radicals, triplet states, compounds with transition metal and rare earth ions and compounds with defect centers. It is a useful technique for analysing orientationally disordered systems such as powders or frozen solutions where standard diffraction methods fail.

Applications

ESR spectroscopy is used to study a wide variety of materials.

1. Inorganic and organic free radicals that have an odd number of electrons—Examples are Fremy's radical, $ON(SO_3)_2^{2-}$ and diphenylpicrylhydrazyl (DPPH). Free radicals are frequently encountered as intermediates in chemical reactions and enzyme–substrate reactions, and organic radicals in proteins, semiquinone radicals.

2. Odd electron molecules such as NO, NO_2, and ClO_2—many such molecules have been examined by gas-phase ESR technique.

3. Triplet-state molecules such as O_2 and S_2—these systems have two unpaired electrons. Optical irradiation of solids and solutions can often

permit investigation of photoexcited triplet states, which are important in photochemistry.

4. Transition metal complexes, organometallics and catalysts containing metal ions with an incomplete 3d, 4d or 5d orbital and metal centres in protein complexes such as Mn(+II), Cu(+II), Mo(V), etc.

5. Rare earth and actinide compounds containing incomplete 4f, 6d or 5f subshells.

6. Impurities in solids, such as semiconductor materials—Odd electrons gained by an acceptor or lost by a donor impurity may be associated with energy bands in crystals.

7. Electrons in the conduction band of metals can be examined by ESR spectroscopy.

Electron Behaviour

The basic physical concepts of this technique are analogous to those of NMR, but instead of the nuclear spins, electron spins are excited. Because of the difference in mass between nuclei and electrons, weaker magnetic fields and higher frequencies are used, compared to NMR. For electrons in a magnetic field of 0.3 tesla, spin resonance occurs at around 10 GHz. The coupling strength is larger by a factor of 10^6 in ESR (MHz instead of Hz) and the relaxation times are 10^6 times smaller in ESR (ns instead of ms). The sensitivity is better by a factor of 10^6 than in NMR (1 nM instead of 1 mM).

An electron has a magnetic moment. When placed in an external magnetic field of strength B_0, this magnetic moment can align itself parallel or antiparallel to the external field. The former is a lower energy state than the latter (this is the Zeeman effect). These two levels are given by

$$E = g_e \mu B_0 M_s$$

where,

M_s is the electron spin quantum number (+1/2 or −1/2),

g_e is the gyromagnetic ratio of the electron (also called the Landé g-factor) and

μ_B is the Bohr magneton.

The electron can absorb electromagnetic radiation having energy equal to the energy difference between the two levels. Then it can move from lower level to higher level.

To move between the two energy levels, the electron can absorb electromagnetic radiation of the correct energy. The energy separation between the two levels is given by

$$\Delta E = h\nu$$

$$= g_e \mu_B B_0$$

and this is the fundamental equation of ESR spectroscopy. The paramagnetic centre is placed in a magnetic field and the electron is allowed to cause resonance between the two states; the energy absorbed as it does so is monitored, and converted into the EPR spectrum (Figure 7.1).

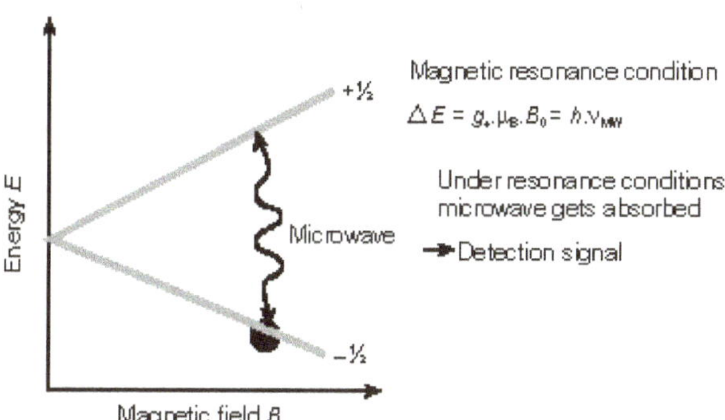

Figure 7.1 Absorption of microwave radiation energy under resonance conditions

A free electron (on its own) has a g-value of 2.002319304386. The electronic g factor is the ratio of magnetic dipole moment to its angular momentum. It is a dimensionless scalar quantity and is called g-factor or spectroscopic splitting factor. Use of the term g-value is discouraged.

For radiation at the commonly used frequency of 9.5 GHz (known as X-band microwave radiation, and thus giving rise to X-band spectra), resonance occurs at a magnetic field of about 0.34 tesla (3400 gauss) (Figure 7.2).

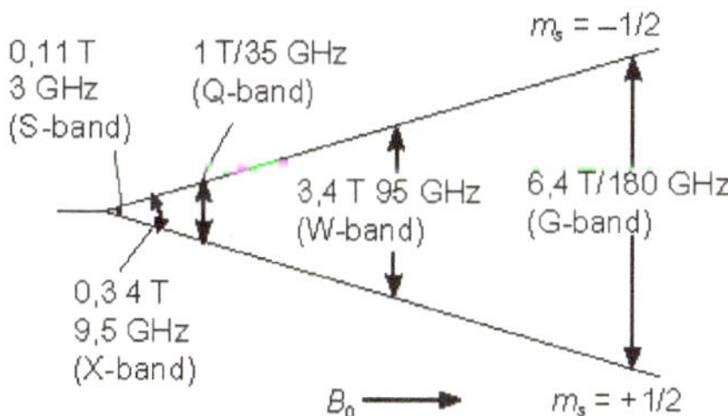

Figure 7.2 Microwave frequency bands

EPR signals can be generated by resonant energy absorption measurements made at different electromagnetic radiation frequencies ν in a constant external magnetic field (i.e., you scan with a range of different frequency radiation whilst holding the field constant, like in an NMR experiment). Conversely, measurements can be provided by changing the magnetic field B and using a constant frequency radiation; due to technical considerations, this second way is more common. This means that an EPR spectrum is normally plotted with the magnetic field along the X-axis, with peaks at the field that cause resonance on the other hand an NMR spectrum has peaks at the frequencies that cause resonance) (Figure 7.3).

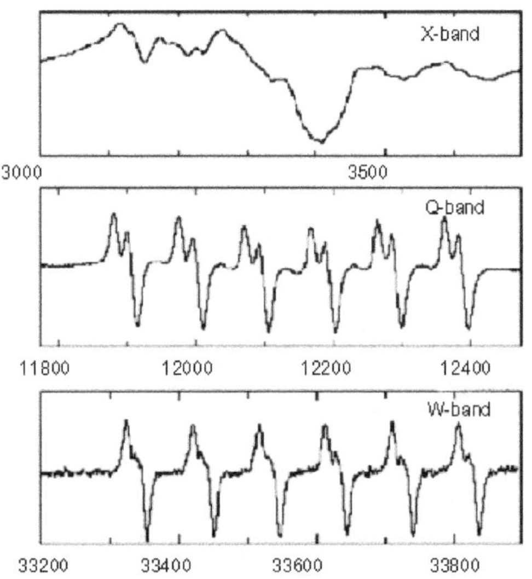

Figure 7.3   Multifrequency-EPR on cytochrome *c* oxidase

Oscillating magnetic field (frequency and amplitude)   The **frequency** (γ) of the oscillating magnetic field applied to induce transitions between the magnetic energy levels of electrons is measured in gigahertz (GHz instead of MHz). This factor is 1000 times larger in EPR than in NMR. The **amplitude** of the oscillating magnetic field is designated as $B1$. The recommended unit is the millitesla (MT).

Static magnetic field   The static magnetic field at which the EPR spectrometer operates is measured by the magnetic flux density B and the recommended unit is the tesla (T) (1 T = 10^4 gauss).

EPR absorption and dispersion   A single transition and a set of degenerate or unresolved transitions are referred to as a line. The line shape is often described to be Lorentzian, Gaussian, or a mixture of the two. Absorption or dispersion lines are commonly presented in the first or the second derivative mode.

In practice a single, isolated, paramagnetic centre never occurs, but only a population of a large number of centres. If this population of centres is in thermodynamic equilibrium, its statistical distribution is described by the Boltzmann ditribution.

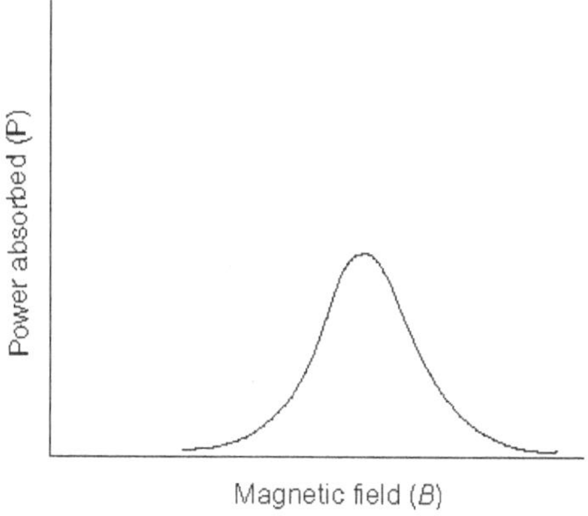

Figure 7.4   ESR absorption peak (power absorbed versus magnetic field)

ESR spectrum can be represented as an absorption spectrum by plotting absorb power against magnetic field (Figure 7.4). However, the first-derivative spectrum is the usual form obtained using ESR spectrometers (Figure 7.5).

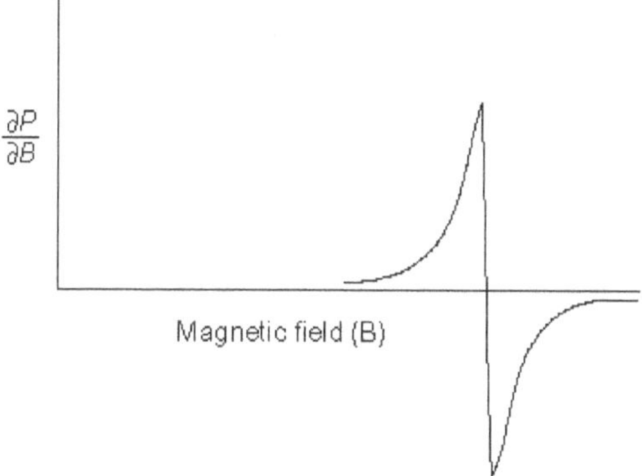

Figure 7.5   ESR first-derivative spectrum

Here rate of change of absorption is plotted against magnetic field strength. In general, ESR absorption are broad, first-derivative spectra enable more accurate measurement.

Instrumentation

ESR instrument should consists of the following components.

1.   Microwave source
2.   Sample

3. Detector

4. Magnet

A simple form of ESR instrument is shown in Figure 7.6. The microwave source is a **klystron** oscillator, delivering 30 to 300 mW of power. The energy is transmitted by means of a **waveguide**, a rectangular tube made of copper or brass with dimensions appropriate to the wavelength of radiation. The sample is placed in a **resonant cavity** which concentrates the source energy for the sample by multiple reflections of the travelling microwave from the two end walls. Detection is accomplished with a semiconducting crystal detector which acts as a **rectifier**, converting the microwave power into a direct current. The magnetic field supplied by an electromagnet is arranged perpendicular to the field of the microwave radiation.

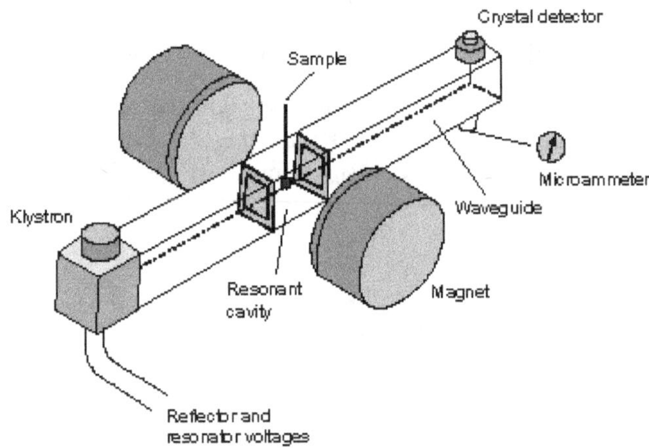

Figure 7.6   Simple form of ESR spectrometer

In operation, the magnetic field strength is slowly varied while a constant microwave source frequency is maintained. As the magnetic field strength approaches the value corresponding to resonance for the sample being studied, power is absorbed by the sample and the power transmitted through the cavity to the detector is reduced. The decrease in measured current represents the ESR signal.

Reflection Cavity in ESR Spectrometer

To increase the sensitivity, modern instruments use a reflection cavity located at right angles to the source and the detector. This arrangement necessitates some means of transmitting the microwave power into the sample cavity and then to the detector. Several types of divider devices are available including a hybrid tee, a circulator and a directional coupler.

The hybrid tee is a device which will not allow microwave power to pass in a straight line from one arm to the opposite. Instead, it turns the corner as shown by the arrows in Figure 7.7. Thus power coming from arm 1 is equally divided between arms 2 and 3. Arm 3 usually contains a resistor, which forms one arm of a balanced Wheatstone bridge circuit. Any change in the impedance of arms 2 or 3 will unbalance the bridge, resulting in a signal reaching the crystal detector. In practice, when resonance conditions for the sample are not satisfied, power will be absorbed in the sample cavity and the impedance of arm 2 will balance that of arm 3. At resonance, however, the sample absorbs energy from the microwave field and this energy change unbalances the impedance of the cavity. This is reflected in the change in signal received by the crystal detector.

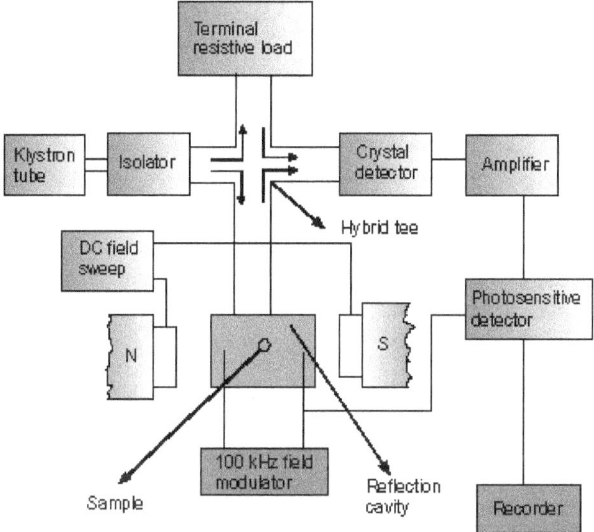

Figure 7.7   Block diagram of a modern ESR spectrometer.

Applications   This reflection cavity arrangement has two advantages over simple transmission cavity.

1. It is more efficient in building up microwave power and so gives twice the sensitivity of transmission cavity.

2. This bridge circuit gives a positive signal only in the region of a resonance peak. No signal is detected off resonance.

Isolator

The isolator allows transmission of microwave radiation in only one direction from the klystron tube to the hybrid tee and inhibits the reverse transmission. In this way, there is no danger that the klystron will become detuned through reflective feedback from the sample cell at resonance.

Samples

Solid, liquid or gaseous samples can be used in ESR spectroscopy. Gaseous samples normally give signals that are too weak to be detected accurately. Samples are usually enclosed in cylindrical quartz tubes since glass often contains a sufficient quantity of paramagnetic impurities to interfere.

Thermal equilibrium and stabilization   For a sample in thermodynamic equilibrium containing N spin systems (S = 1/2) in a magnetic field, there is a population difference between the two M_s levels arising from each S = 1/2 system. The difference follows the Boltzmann distribution.

$$\frac{NMs = +1/2}{MMs = -1/2} = e - hv / kT \approx 1 - hv / kT$$

In this expression, NM_s = ± 1/2 is the number of spins having M_s = +1/2 and M_s = −1/2. Under normal conditions, there is a slight excess population in the lower M_s = −1/2 level. Absorption of microwave energy by the sample induces transition from the M_s = −1/2 to M_s = +1/2 level. To maintain steady-state conditions, electrons promoted to the excited state must lose energy and return to the lower level. Otherwise, saturation would occur and no resonance absorption would be observed. This is similar to the situation in NMR spectroscopy.

Saturation is normally avoided by working at low RF power levels. The electrons lose energy and return to the ground state by two relaxation mechanisms: spin–lattice and spin–spin relaxation.

ESR spectra

In a homogeneous magnetic field, an unpaired electron in an assemblage of other atoms can have a number of energy states. Since the radical electron is usually delocalized over the whole molecule or at least a large part of it, the unpaired electron comes into contact with many nuclei. Nuclei possessing magnetic moment may interact and cause a further splitting of the electron resonance line.

The selection rules for allowed ESR transitions are

$$\Delta M_I = 0 \text{ and } \Delta M_S = \pm 1$$

The simplest free radical is the hydrogen atom. The energy levels and observed ESR spectrum of hydrogen are depicte in Figure 7.8.

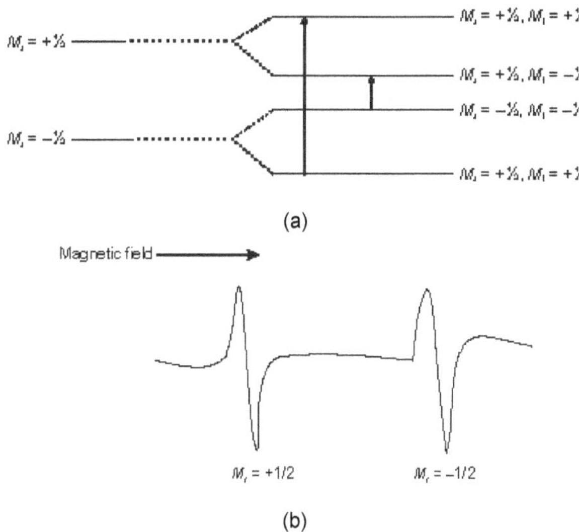

(a)

(b)

Figure 7.8   (a) Energy levels for the hydrogen atom (b) ESR spectrum of hydrogen atom

The nuclear spin ($I = 1/2$) interacts with the electron spin and with the external magnetic field. The result of these interactions is that the ESR spectrum consists of two peaks.

For an ion with a non-zero nuclear moment I, $2I+1$ lines can be expected centered around g. For example, for ^{63}Cu(II) (a $3d^9$ ion), $I = 3/2$. Four lines are expected and generally observed in the first-derivative spectrum.

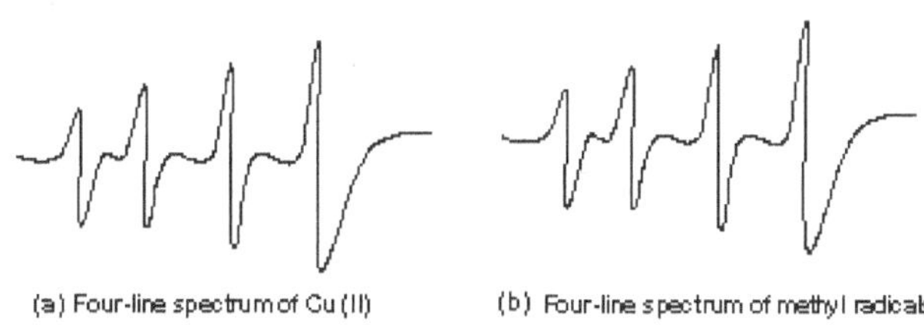

Figure 7.9   ESR spectrum (*Continues*)

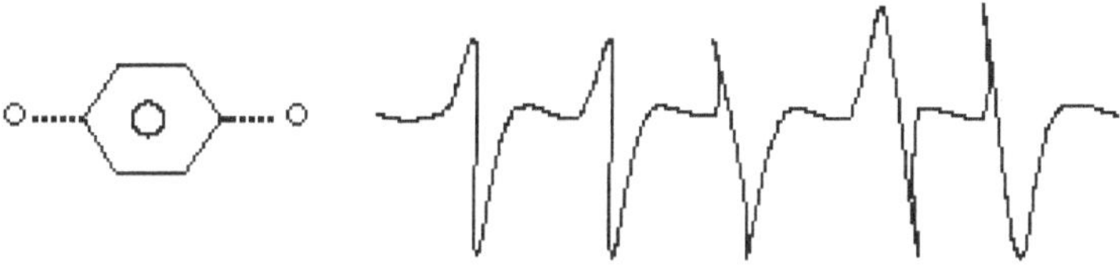

Figure 7.9   ESR spectrum

Hyperfine Coupling

Multiplet structure of ESR spectra Quite often, ESR spectrum is not very simple but it exhibits multiple structures. There are two kinds of multiplet structures.

Fine structure It generally involves large energy separations between multiplet lines. This type is of limited importance because it is observed only in solids and in cases where paramagnetic material has more than one unpaired electron. For example, molecular oxygen has two unpaired electrons and thus it is a stable triplet state molecule. The unpaired electrons in oxygen are so also that the spin–spin splitting is larger than applied magnetic field. So fine splitting will not be observed with oxygen under normal resonance conditions.

Hyperfine structure These are analogous to spin–spin coupling in NMR. It involves much smaller energy separations. It arises through coupling of an unpaired electron with neighbouring nuclear spins. When the nuclei are chemically equivalent, the number of lines is given by

$$2n_i I + 1$$

where n_i = number of equivalent nuclei having nuclear spin I.

For n equivalent nuclei with $I = 1/2$, there will be $n + 1$ lines. These lines will have intensities proportional to the binomial expansion of order n.

Organic free radicals in solution generally exhibit hyperfine coupling with several nuclei. Thus methyl radical (CH_3) exhibits a four-line spectrum from splitting by three equivalent hydrogens.

The p-benzosemiquinone radical anion gives a five-line spectrum. The above-mentioned four-line and five-line spectrum are shown in Figure 7.9.

Interpretation of ESR Spectra

The interpretation of ESR spectra involves several parameters: the g factor, the separation of the hyperfine lines and their relative intensities, the sample concentration, relaxation times and linewidths. Frequently the measured spectrum will not contain all the lines expected because the g factor and coupling constants may be such that two lines come close to one another and are not resolved. When many equivalent nuclei interact, relative peak heights become very large and it is difficult to see the smallest peaks. These smaller peaks are important because the outer portions of a spectrum are invariably simplest. Some examples of interpreting an ESR spectrum are given below.

Example 1

The ESR spectrum of the ethyl radical trapped in argon at 4.2 K consists of a quartet of relatively sharp lines. Each of these lines is further split into a triplet. The outer lines of the triplet are relatively broad. In liquid ethane at $-170°C$, the spectrum consists of 12 very sharp lines.

Interpretation   The quartet is due to CH_3 and indicates that the unpaired electron interacts equally with all three hydrogen atoms and therefore CH_3 freely rotates about the C—C bond. The triplet is caused by the CH_2 protons which produce a dipole magnetic anisotropy which cannot be averaged out by internal rotation. End-over-end rotation does not take place at 4.2 K. At higher temperatures, rapid tumbling of radicals in solution cancels the anisotropic effects.

When the organic radical contains two or more sets of equivalent atoms, the total number of lines in ESR spectrum is given by

$$\prod_{I=1}^{j}(2n_i\,I_i + 1) = (2n_1 I_1 + 1)(2n_2 I_2 + 1)...(2n_j I_j + 1)$$

Therefore, for ethyl radical,

$$(2 \times 3 \times 1/2 + 1)\,(2 \times 2 \times 1/2 + 1)\ = 4 \times 3 = 12$$

Thus 12 lines are obtained.

Example 2

Predict the type of ESR spectrum to be obtained for 2, 3-dichlorobenzoquinone.

Since the only nuclei of non-zero spin are the two protons and since they are equivalent, the ESR spectrum should be triplet of relative intensity 1:2:1.

Electron–electron interaction

We have so far considered radicals which have just one unpaired electron. However, there are many systems which have several unpaired electrons and these may also be studied by ESR. For example, molecules having diamagnetic ground states may have excited triplet states which have different lifetimes long enough for their ESR spectra to be recorded.

In a magnetic field, a triplet state splits into its three components giving two possible transitions,

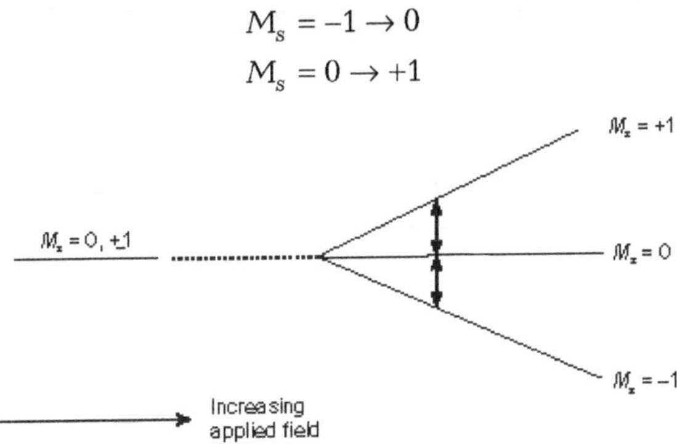

Figure 7.10   Splitting of a triplet state in an applied magnetic field

If two unpaired electrons occupy sites which are well-separated in the molecule, little or no interaction occurs between the two electrons and each gives an electron resonance signal though it belonged to a separate monoradical, i.e., no splitting occurs and a single peak is observed. If two electrons are sufficiently close, as in molecules in the triplet state, interactions become important. Often the interaction between electrons is so strong that it overshadows the effect of the externally applied magnetic field. For example, molecular oxygen has two unpaired electrons in its ground state and is thus a triplet molecule. However, the unpaired electrons in oxygen are so close together that spin–spin splitting is larger than the external magnetic field. This essentially means that the electrons will precess about the molecular axis rather than about the external magnetic field. Thus line splitting will not be observed with oxygen under normal resonance conditions.

Kramer's Rule

Kramer's rule states that for a system with an even number of unpaired electron spins, the lowest energy state will be that with $M_s = 0$ as indicated in the case of triplet state. All higher energy states in this triplet and all states for systems with an odd number of unpaired spins, will be doubly degenerate corresponding to the two signs possible for $M_s = 0$. This degeneracy is known as Kramer's degeneracy. It is removed by the applied field which shifts the two levels $M_s = -1$ and $M_s = +1$ in opposite directions.

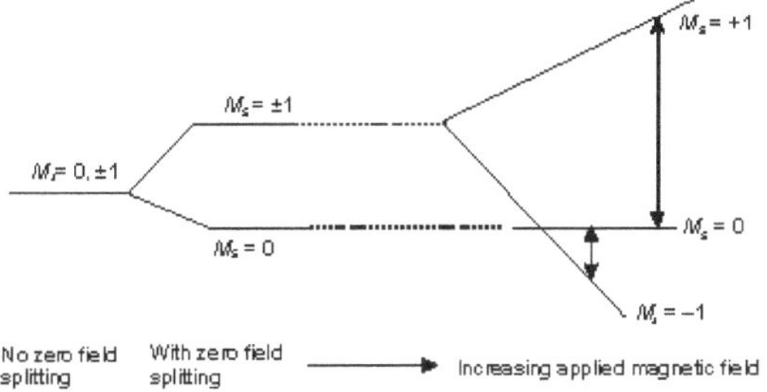

Figure 7.11   Effect of magnetic field on the energy levels of a triplet state which exhibits zero-field splitting

The important consequence is that the two allowed transitions $M_S = -1 \to 0$ and $M_S = 0 \to +1$) which were of equal energy in the absence of zero-field splitting, are now of quite different energy and two signals are observed.

Modern Pulse EPR

Traditional continuous wave EPR methods are quite limited in spectral and time resolution, thus the vast potential of EPR spectroscopy cannot be fully utilized. This is in contrast to modern pulse EPR, a new branch of electron paramagnetic resonance, which allows for the design of experiments that can specifically address a given problem and provide utmost resolution by separating interactions from each other. Recently, a multitude of pulse EPR experiments for increasing spectral resolution and sensitivity have been developed. Some of the methods that can be used to achieve detailed information about a paramagnetic species are schematically illustrated in Figure 7.12 .

Field-swept EPR techniques reveal information about the electronic state of the material under study (electron Zeeman interaction), about the coupling between unpaired electrons (fine structure term) and about strong interactions between the unpaired electron and the surrounding nuclei (hyperfine couplings) (Figure 7.12a). Interactions with more distant nuclei can be investigated by means of electron nuclear double resonance (ENDOR) (Figure 7.12b), and electron spin echo envelope modulation (ESEEM) and hyperfine correlation (HYSCORE) spectroscopy (Figure 7.12c).

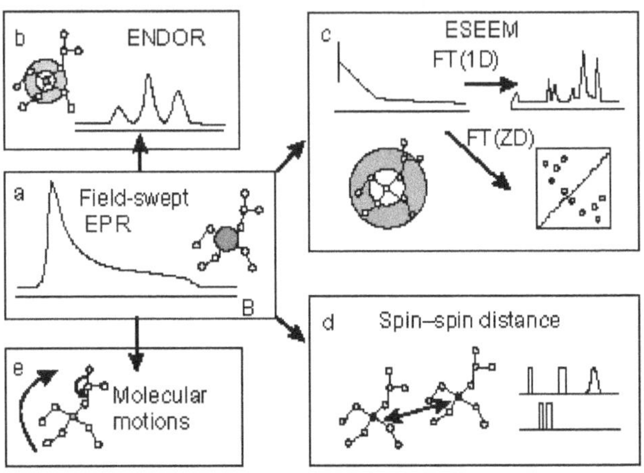

Figure 7.12   Modern pulse EPR methods

With these techniques the spectral resolution can be improved by orders of magnitude as compared to standard EPR methods. ENDOR, ESEEM and HYSCORE spectra contain information on the type of nuclei in the vicinity of the unpaired electron (nuclear Zeeman interactions), the distances between the nuclei and the electron spin and the spin density distribution (hyperfine interactions), and the electric field gradient caused by the electrostatic charges (nuclear quadrupole interactions). Furthermore, a number of pulse techniques have been invented to study distances between different paramagnetic centres (Figure 7.12d), and to investigate dynamic properties of paramagnetic molecules (Figure 7.12e). Pulse EPR methods can thus provide detailed local structural information of ordered and disordered paramagnetic compounds.

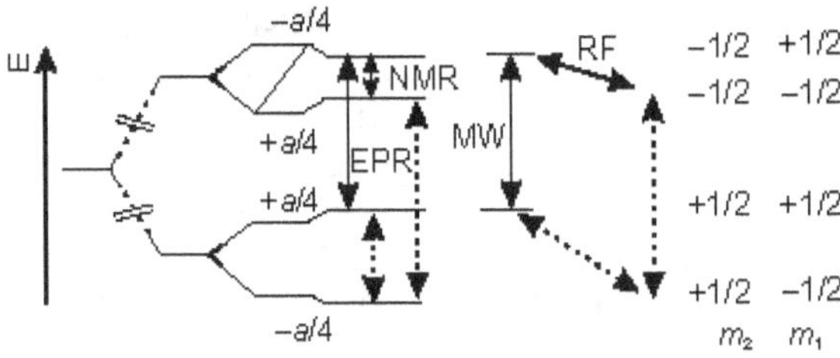

Figure 7.13   Electron nuclear double resonance (ENDOR) (Anisotropic hyperfine interactions)

In ENDOR, a sample is irradiated simultaneously with a microwave frequency suitable for electron resonance and a radio frequency suitable for nuclear resonance. In the Figure 7.13, microwave frequency (MW) helps to measure the transition between $+a/4$ and $-a/4$, i.e., M_2 values $+1/2$ to $-1/2$. Simultaneously broadening of electron resonance line by nuclear energy level is avoided.

ELDOR

Consider a system which has a single electron spin and a nucleus with spin ½. There are two allowed electron spin transitions, one of which is observed while the other is irradiated. As these transitions do not share energy levels, the effects of irradiation (also called pumping) are only seen as a result of spontaneous nuclear spin transitions. In an ELDOR spectrum, the intensity change in the monitored resonance is plotted against frequency difference. For a more complicated system, the spectrum may have several lines, each single line giving a direct measurement of splittings. ELDOR spectra also contain lines corresponding to pumping of the forbidden transitions.

QUANTITATIVE ANALYSIS

The integrated intensity is usually related to the concentration of the paramagnetic species in comparison with a standard. The total area enclosed by either the absorption or dispersion signal is proportional to the number of unpaired electron spins in the sample. Comparison is made with a standard containing a known number of unpaired electrons and having the same line shape as the unknown. A solid frequently used is 1, 1-diphenyl-2'-picrylhydrazyl (DPPH) or solutions of peroxylamine disulphonate. DPPH contains 1.53×10^{23} unpaired spins per gram. Secondary standards include charred dextrose or synthetic ruby attached top the cavity. A dual sample cavity is used to minimize difficulties with actual physical interchange of standard and sample.

PROBLEMS AND SOLUTIONS

1. How many ESR lines can be expected for the 1,2 -,1,3- and 1,4-difluorobutadiene radical anions? $I = 1/2$ for ^{19}F and ^{1}H.

    Solution

    64 lines for 1, 2- and 1, 3-compounds (no hydrogen atoms or fluorine atoms are equivalent); 27 or 64 lines for the 1, 4 compound depending on whether the Fs are *trans–trans* or *trans–cis* substituted.

2. How many ESR lines could be expected for the ^{33}S^{19}F$_6$ radical anion? $I = 3/2$ for ^{33}S and $I = 1/2$ for ^{19}F.

    Solution

    Number of lines = $(2 \times 1 \times 3/2 + 1)(2 \times 6 \times 1/2+1) = 4 \times 7 = 28$ lines.

3. How many ESR lines can be expected for the $Cu(S_2PF_2)_2$ and $V(S_2PF_2)_2$ if the metal, phosphorus and fluorine hyperfine splittings are observable in the solution spectra ? $I = 3/2$ for Cu and S; $I = 7/2$ for V; $I = 1/2$ for F and P.

Solution

60 lines for copper compound and 120 lines for the vanadium compound.

4. Which valency states of copper and silver will show a strong ESR signal?

Solution

Copper (II) and silver (II)

5. When malonic acid, $CH_2(COOH)_2$, is irradiated with X-rays at room temperature, the ESR spectrum appears to consist of a dominant doublet and a less intense overlapping triplet. On standing for a few days, only a doublet remains. Determine the two products that are formed.

Solution

The predominant radical is $CH(COOH)_2$ with a small amount of CH_2COOH.

REVIEW QUESTIONS

1. Which valency states of copper and silver will show strong ESR signals?

2. Calculate the static magnetic field required and the value of ΔE for an ESR spectrometer operating at a frequency of 35,000 MHz.

3. Predict the ESR spectrum of benzene radical anion.
   (Ans: 7 line signal)

4. The ESR signal of radical produced by abstraction of H from diethyl ether shows a complex hyperfine structure. Give reasons.

5. What are fine structures and hyper fine structures in ESR spectra?

6. Explain the instrumentation involved in a simple ESR spectrometer.

7. Mention the selection rule of ESR spectroscopy.

8. What is zero-field splitting?

9. What is meant by Kramer's degeneracy?

10. C12 nuclei do not interact with the electron. Why?

11. Which of the following will show ESR spectra and why?

    i. N_2 ii. H_2 iii. O_2 iv. Cu^{2+} v. Cu^+ vi. H

12. What is the principle of ESR?

13. Why are water and alcohol not suitable solvents for ESR studies?

14. Show how ESR derivative spectra having three lines can arise by

    i. hyperfine coupling to two nuclei with spin 1/2

    ii. hyperfine coupling to a nucleus with spin 1.

    How can these possibilities be distinguished?

Chapter 8

MÖSSBAUER SPECTROSCOPY

Introduction

Mössbauer spectroscopy involves transitions within the nucleus. It deals with the emission and absorption of gamma rays. The technique is sometimes called nuclear gamma resonance spectroscopy. The only suitable sources of radiation are excited nuclei of the same isotope formed during radioactive decay. There is no direct way of tuning the emitted quanta to the energy needed to excite the absorber. The necessary tuning, which is essential with monochromatic sources, is achieved using the Doppler effect by introducing controlled relative motion of source and absorber. This explains why the energy differences in Mössbauer spectroscopy are recorded in unfamiliar units of mm s^{-1}. Mössbauer spectroscopy is a versatile technique that is useful in many areas of science such as physics, chemistry, biology and metallurgy. It can give very precise information about the chemical, structural, magnetic and time-dependent properties of a material. The key to the success of the technique is the discovery of recoilless gamma ray emission and absorption, now referred to as the 'Mössbauer effect', after its discoverer Rudolph Mössbauer who first observed the effect in 1957 and received the Nobel Prize in Physics in 1961 for his work.

Principle

The Mössbauer Effect

The nuclei in atoms undergo a variety of energy level transitions, often associated with the emission or absorption of a gamma ray. These energy levels are influenced by their surrounding environment, both electronic and magnetic, which either change or split energy levels. The changes in energy levels can provide information about the atom's local environment within a system when observed using resonance-fluorescence. There are, however, two major obstacles in obtaining this information: (a) the 'hyperfine' interactions between the nucleus and its environment are extremely small and (b) the recoil of the nucleus as the gamma ray is emitted or absorbed, which prevents resonance.

A free nucleus after emission or absorption of a gamma ray recoils due to conservation of momentum, just like how a gun recoils after firing a bullet, with a recoil energy E_R. This recoil is shown in Figure 8.1. The emitted gamma ray has E_R less than that of the nuclear transition, but to be resonantly absorbed it must be greater than the transition energy due to the recoil of the absorbing nucleus. To achieve resonance, the loss of recoil energy must be overcome in some way

Figure 8.1   Recoil of free nuclei in emission or absorption of a gamma ray

As the atoms move due to random thermal motion, the gamma ray energy has a spread of E_D caused by the Doppler effect. This produces a gamma ray energy profile as shown in Figure 8.2. To produce a resonant signal, the two energies need to overlap and this is shown in the shaded area. This area is shown exaggerated, because in reality, it is extremely small, with a millionth or less of the γ-rays in this region.

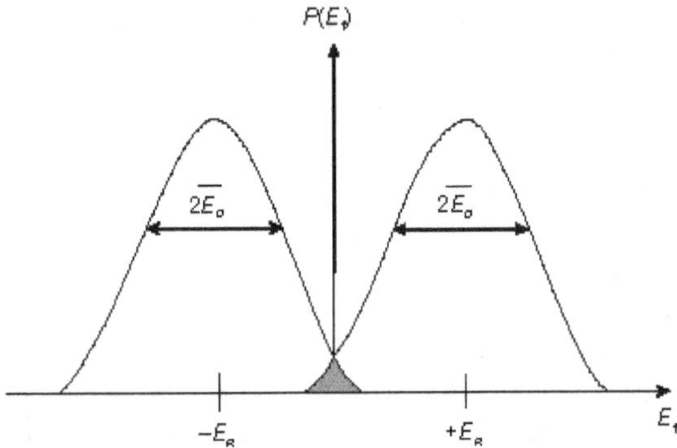

Figure 8.2   Resonant overlap in free atoms. The overlap (shaded region) is greatly exaggerated.

Mössbauer discovered that, when the atoms are within a solid matrix, the effective mass of the nucleus is very much greater. The recoiling mass is now effectively the mass of the whole system, making E_R and E_D very small. If the gamma ray energy is small enough, the recoil of the nucleus is too low to be transmitted as a phonon (vibration in the crystal lattice), and so the whole system recoils, making the recoil energy practically zero: a recoil-free event. In this situation, as shown in Figure 8.3, if the emitting and absorbing nuclei are in a solid matrix, the energies emitted and absorbed gamma ray will be the same (i.e., resonance).

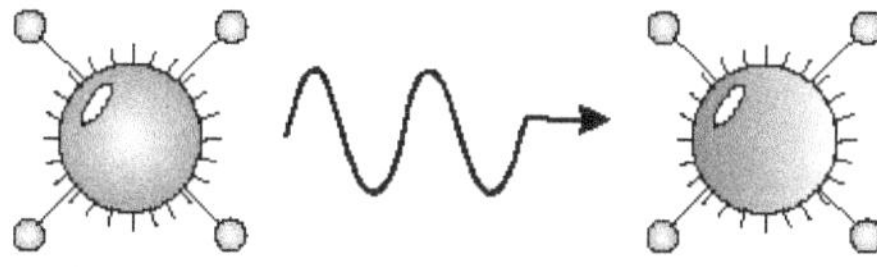

Figure 8.3   Recoil-free emission or absorption of a gamma ray when the nuclei are in a solid matrix such as a crystal lattice

If the emitting and absorbing nuclei are in identical, cubic environments, then the transition energies are identical. This produces a spectrum as shown in Figure 8.4, which is a single absorption line.

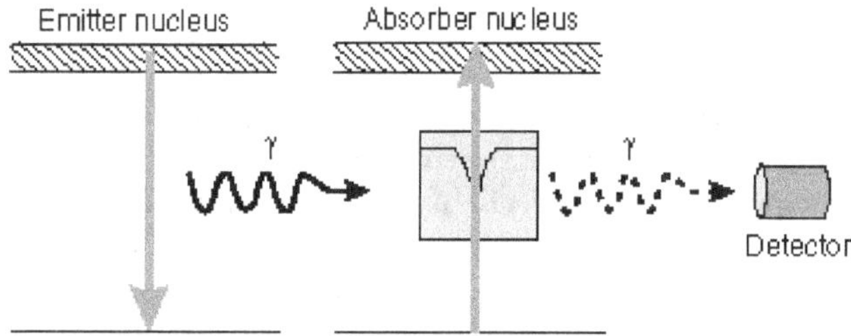

Figure 8.4   Simple Mössbauer spectrum from identical source and absorber

Now that we can achieve resonant emission and absorption, Is it possible to probe the tiny hyperfine interactions between an atom's nucleus and its environment? The limiting resolution, now that recoil and Doppler broadening have been eliminated, is the natural linewidth of the excited nuclear state. This is related to the average lifetime of the excited state before it decays by emitting the gamma ray. For the most common Mössbauer isotope, ^{57}Fe, this linewidth is 5×10^{-9} eV. As compared to the Mössbauer gamma ray energy of 14.4 keV, this gives a resolution of 1 in 10^{12}, or the equivalent of one sheet of paper in the distance between the sun and the earth. This exceptional resolution is of the order necessary to detect the hyperfine interactions in the nucleus.

As resonance occurs only when the transition energy of the emitting and absorbing nucleus match exactly, the effect is isotope-specific. The relative number of recoil-free events (and hence the strength of the signal) is strongly dependent upon the gamma ray energy, and so the Mössbauer effect is detected only in isotopes with very low-lying excited states. Similarly, the resolution is dependent upon the lifetime of the excited state. These two factors limit the number of isotopes that can be used successfully for Mössbauer spectroscopy. The most widely used is ^{57}Fe, which has both a very low-energy gamma ray and long-lived excited state. Figure 8.5 shows the isotopes in which the Mössbauer effect has been detecte.

H																	He
Li	Be											B	C	N	O	F	Ne
Na	Mg											Al	Si	P	S	Cl	Ar
K	Ca	Sc	Ti	V	Cr	Mn	Fe	Co	Ni	Cu	Zn	Ga	Ge	As	Se	Br	Kr
Rb	Sr	Y	Zr	Nb	Mo	Tc	Ru	Rh	Pd	Ag	Cd	In	Sn	Sb	Te	I	Xe
Cs	Ba	Lu	Hf	Ta	W	Re	Os	Ir	Pt	Au	Hg	Tl	Pb	Bi	Po	At	Rn
Fr	Ra	Ac															

Ce	Pr	Nd	Pm	Sm	Eu	Gd	Tb	Dy	Ho	Er	Tm	Yb	Lu
Th	Pa	U	Np	Pu	Am	Cm	Bk	Cf	Es	Fm	Md	No	Lr

Figure 8.5   Elements of the periodic table that have known Mössbauer isotopes (shown in grey); those used the most are shaded black.

The Mössbauer effect is the recoil-free emission of gamma radiation from a solid radioactive material. Since the gamma emission is recoil-free, it can be resonantly absorbed by stationary atoms, i.e., also in a solid. The nuclear transitions are very sensitive to the local environment of the atom, and Mössbauer spectroscopy is a sensitive probe of the different environments an atom occupies in a solid material.

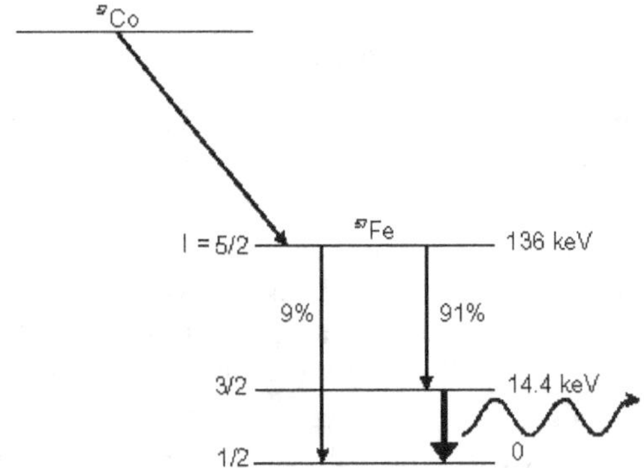

Figure 8.6   Production of gamma rays for ^{57}Fe Mössbauer spectroscopy

Approximately 90% of the ^{57}Fe nuclear excited state decays through the intermediate level to produce 14.4 keV gamma radiation. These gamma photons can then be absorbed by ^{57}Fe in a sample (Figure 8.6).

Conditions for Mössbauer Spectroscopy

Some of the most important conditions to be satisfied are the following.

1.   The energy of the nuclear transition must be large enough to give useful gamma radiation, but not large enough to produce a recoil whose energy is much greater than the lattice vibrations quanta.

2. A substantial proportion of the excited nuclei must decay with emission of gamma photons. Other decay processes, such as the emission of conversion electrons, must not dominate.

3. The lifetime in the excited state must be long enough to avoid introducing too much uncertainty in E, through the uncertainty principle. It must be short enough to give intense lines of reasonable breadth. In practice, lifetimes of 1 and 100 ns are suitable.

4. The excited state of the emitter must have a precursor which is long lived and reasonably easy to handle.

5. The ground state isotope must be stable. It should be present in fairly high natural abundance or enrichment should be easy.

6. The cross section for absorption should be high.

Instrumentation

The gamma ray source is a radioactive element that is mechanically vibrated back and forth to produce a Doppler shift in the energy of the emitted gamma radiation. Figure 8.7 shows a transmission Mössbauer experiment. As the energy of the gamma radiation is scanned by Doppler shifting, the detector records the frequencies of gamma radiation that are absorbed by the sample

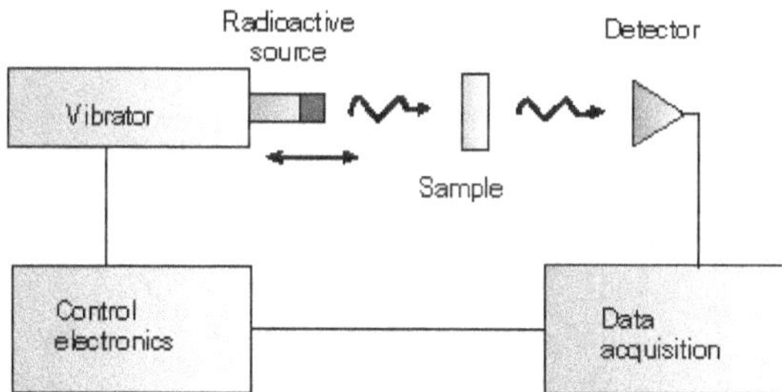

Figure 8.7   Experimental set-up for transmission Mössbauer spectroscopy

The Mössbauer spectra are usually recorded in the transmission geometry using a constant acceleration spectrometer operated in connection with a multi-channel analyser in the time-scale mode. The source is kept at room temperature and consists of approximately 20 mCi of Co^{57} diffused into rhodium or paladium foil. The spectrometer is calibrated against a metallic iron foil and zero velocity is taken as the centroid of its room temperature Mössbauer spectrum. In such calibration spectra, linewidths of about 0.23 mm/s were normally observed. The duration of a typical run is a few hours for non-biological compounds and 24 hours for proteins. A typical spectrometer is depicted in Figure 8.8.

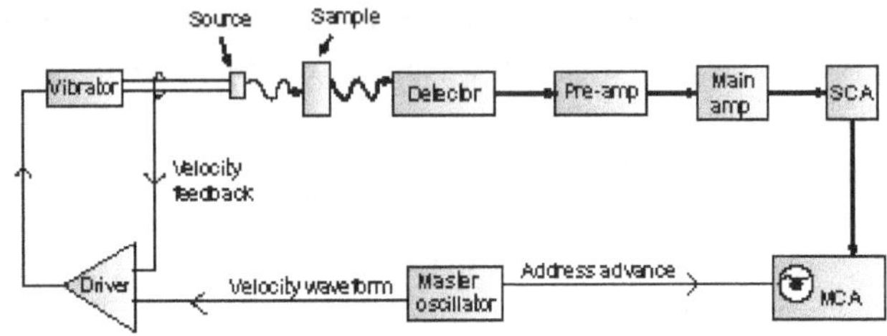

Figure 8.8   Schematic representation of the Mössbauer spectrometer

As shown in Figure 8.8, both the source velocity and the address of the active channel of the multichannel analyser (MCA) are controlled by the master oscillator. The oscillator synchronizes the source acceleration and the sweep of the memory registers, causing the active channel address to be a linear function of velocity. In other words, the pulses counted, while the source is at a particular velocity, are always stored in a particular register.

The master oscillator output is a rounded saw tooth or triangular wave and is applied to one of the driver inputs. The driver is essentially a difference amplifier with its output applied to the vibrator. A pick-up coil on the vibrator supplies a voltage proportional to the velocity of the source and is connected to the second input of the driver. The driver and the vibrator form a tightly coupled, electro-mechanical, negative feedback loop, forcing the source velocity to be directly proportional to the master oscillator signal. The desired velocity range is selected by scaling the master oscillator signal before it is applied to the driver.

The gas proportional counter contains primarily krypton or argon plus a small amount of methane quench gas. A 14.4-keV photon entering the chamber through the mylar window ionizes some of the Ar or Kr atoms, the creation of each ion/electron pair requiring about 30 eV. The positive ions drift to the outer wall and the electrons to the centre wire. The accelerating electrons ionize more atoms, the effective gas gain being approximately 1000. Methane suppresses the random motion of electrons, thus shortening the travel time to the centre wire and the rise time of the signal pulse. The overall efficiency of a Harwell Argon counter is 60% at 14.4 keV, and a 14.4-keV photon will cause about 10^6 ion pairs to be created. The 122-keV and 136-keV photons entering the counter produce a broad background of lower energy signals. This is called Compton background. The single channel analyser (SCA) discriminates against most of these, but about 20% of the pulses falling in the 14.4 keV window of the SCA are from this Compton background.

The charge pulse from the counter is amplified and differentiated by the pre-amp, which is mounted as close to the counter as possible. Further amplification of the pulse is provided by the main amplifier and its output is applied to SCA. The SCA is set to discriminate against the non-14.4-keV signals. The signals accepted by the SCA are added to the active channel of the MCA.

The observed linewidths are larger than the intrinsic width of 0.19 mm/s since some broadening is always introduced by inhomogeneity of the environment of iron in the source and sample. Additional broadening arises from imperfections in the source driving mechanism. Even so, typical linewidths obtainable in the lab are 0.23 mm/s, which is close to the minimum predicted by the uncertainty principle.

So far, we have seen one Mössbauer spectrum: a single line corresponding to the emitting and absorbing nuclei being in identical environments. As the environment of the nuclei in a system will almost certainly be different to our source, the hyperfine interactions between the nucleus and its environment will change the energy of the nuclear transition. To detect this, we need to change the energy of our probing gamma rays. This section will show how this is achieved and the three main ways in which the energy levels are changed and their effects on the spectrum.

Fundamentals of Mössbauer Spectroscopy

As shown earlier, the energy changes caused by hyperfine interactions are very small, of the order of billionths of an electron volt. Such minuscule variations of the original gamma ray are quite easy to achieve by the use of the Doppler effect. The pitch of ambulance's siren is raised in pitch when it's moving towards you and lowered when moving away from you. Similarly, our gamma ray source can be moved towards and away from the absorber. This is most often achieved by oscillating a radioactive source with a velocity of a few mm/s and recording the spectrum in discrete velocity steps. Fractions of mm/s compared to the speed of light (3×10^{11} mm/s) give the minute energy shifts necessary to observe the hyperfine interactions. For convenience, the energy scale of a Mössbauer spectrum is quoted in terms of source velocity, as shown in Figure 8.9.

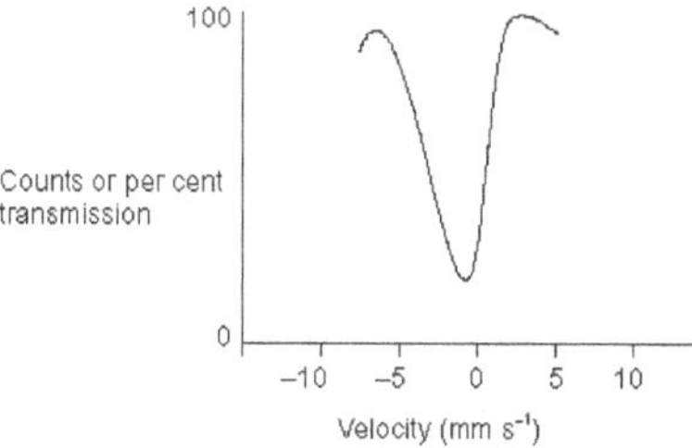

Figure 8.9   Simple spectrum showing the velocity scale and motion of the source relative to the absorber

With an oscillating source, we can now modulate the energy of the gamma ray in very small increments. Where the modulated gamma ray energy precisely matches the energy of a nuclear transition in the absorber, the gamma rays are resonantly absorbed and a peak is produced. The sample must be sufficiently thin to allow the gamma rays to pass through, so that the transmitted ray can be recorded. The relatively low-energy gamma rays can be easily attenuated.

In Figure 8.9, the absorption peak occurs at 0 mm/s, where the source and absorber are identical.

The energy levels in the absorbing nuclei can be modified by their environment in three main ways: isomer shift, quadrupole splitting and magnetic splitting.

Chemical Isomer Shift

Isomer shift arises because the size of the nucleus is different in the ground and the excited states. Isomer shift also called centre shift depends on the electron density at the nucleus. Since S electron wave functions have their maxima at the nucleus, the isomer shift is sensitive to changes in s-electron orbital occupancy. This leads to a monopole (Coulomb) interaction, altering the nuclear energy levels. Any difference in the s-electron environment between the source and absorber produces a shift in the resonance energy of the transition. This shifts the whole spectrum positively or negatively depending upon the s-electron density and sets the centroid of the spectrum.

In a non-relativistic case, the isomer shift is given by the following formula

$$\delta = \left(\frac{Z \cdot e^2 \cdot R^2 \cdot c}{5 \cdot \varepsilon_0 \cdot E_\gamma} \right) \cdot [\rho_a(0) - \rho_s(0)] \cdot \left[\frac{\Delta R}{R} \right] \text{mm s}^{-1}$$

where,

Z is the atomic number,

e is the electronic charge,

R is the effective nuclear radius,

c is the velocity of light,

E_g is the energy of the Mössbauer gamma ray, the $\rho(0)$ terms are the total electron densities at the nucleus for the absorber and source, respectively, Δnd $R = R_{\text{excited}} - R_{\text{ground}}$.

This isomer shift can be calculated and it gives a way of distinguishing between, for example, different ion charge states as they give different isomer shifts. This is due to the different electronic structures of different types of ion.

If the chemical environment of the iron nuclei in the source and absorber are different, then the electron densities will, in general, be different. Since the electromagnetic interaction between the electrons and the nucleus depends, albeit only slightly, on the electron density at the nucleus and on the nuclear radius, and since the radius of the iron nucleus changes slightly in the 14.4 keV transition, there is, in general, a shift of the resonance energy from the source to absorber if the host materials are different. This is called the "isomer shift" because the excited states of nuclei were originally called isomers. The magnitude of the shift depends on the s-electron density at the nucleus. If there is greater electron density at the absorbing nucleus than at the emitting nucleus, additional energy must be given to the gamma ray by moving the source towards the absorber. There are differences on the order of a factor of ten between the isomeric shifts of Fe in the +2 and +3 ionization states in the lattice, and these differences may easily be measured. An isomer shift in a Zeeman or quadrupole pattern appears as a shift of the centre of gravity of the lines.

As the shift cannot be measured directly, it is quoted relative to a known absorber. For example, ^{57}Fe Mössbauer spectra will often be quoted relative to alpha iron at room temperature.

The isomer shift is useful for determining valency states, ligand bonding states, electron shielding and the electron-drawing power of electronegative groups. For example, the electron configurations for Fe^{2+} and Fe^{3+} are $(3d)^6$ and $(3d)^5$, respectively. The ferrous ions have less s-electrons at the nucleus due to the greater screening of the d-electrons. Thus, ferrous ions have larger positive isomer shifts than ferric ions.

Quadrupole Splitting

Nuclei with an angular momentum quantum number I > 1/2 have a non-spherical charge distribution. This produces a nuclear quadrupole moment. In the presence of an asymmetrical electric field (produced by an asymmetric electronic charge distribution or ligand arrangement), this splits the nuclear energy levels. The charge distribution is characterized by a single quantity called the electric field gradient (EFG).

As the quadrupole moment is fixed for each state/isotope, this can be used to probe electronic configuration.

This interaction can be used as a guiding point for fitting the spectra. The largest component of EFG is given by the formula

$$V_{zz} = \frac{1}{4\pi\varepsilon_0} \sum_i q_i r_i^{-3} (3\cos^2\theta_i - 1)$$

and is associated with the quadrupole splitting of energy levels by the following formula

$$\Delta = eQV_{zz}$$

The angular dependance of quadrupole splitting can be used to determine the orientation of spins in the structures being observed. For example, ^{57}Fe has I = 3/2 and leads to two states $m_1 = 1/2$ and $m_2 = 3/2$ (Figure 8.1).

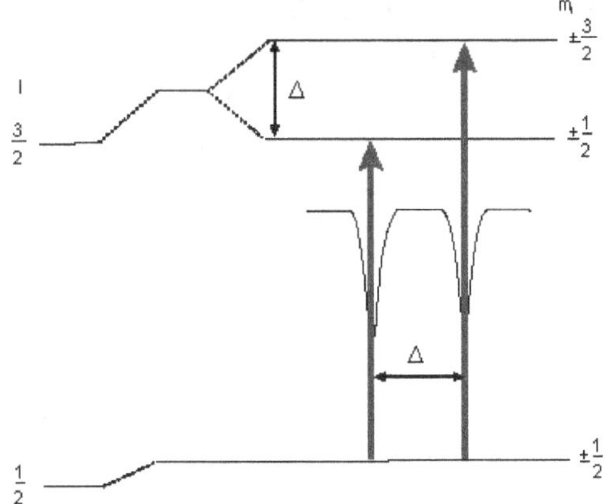

Figure 8.10 Quadrupole splitting for a 3/2 to 1/2 transition. The magnitude (D) of quadrupole splitting is shown.

The nucleus may be in an electric field with a gradient due to the nearby ions with unfilled atomic levels. If the nucleus has a quadrupole (or higher) electric moment, i.e., if its electric charge distribution is not spherically symmetric, its energy levels will be split by amounts that depend on the projections of its spin in the direction of the field gradient. The ground state is shifted by quadrupole interaction, and the first excited level is split into two levels; thus, two transitions are possible, yielding two absorption peaks.

In the case of an isotope with I = 3/2 excited state, such as ^{57}Fe or ^{119}Sn, the excited state is split into two substates m_I = ±1/2 and m_I = ±3/2. This is shown in Figure 8.10, giving a two-line spectrum or doublet.

The magnitude of splitting (D) is related to the nuclear quadrupole moment, Q, and the principle component of the EFG, V_{zz}, by the relation $D = eQV_{zz}/2$.

Magnetic Splitting or Zeeman Splitting

In the presence of a magnetic field, the nuclear spin moment experiences a dipolar interaction with the magnetic field called Zeeman splitting. There are many sources of magnetic fields that can be experienced by the nucleus. The total effective magnetic field at the nucleus, B_{eff}, is given by:

$$B_{eff} = (B_{contact} + B_{orbital} + B_{dipolar}) + B_{applied}$$

The first three terms are due to the atom's own partially filled electron shells. $B_{contact}$ is due to the spin on those electrons polarizing the spin density at the nucleus; $B_{orbital}$ is due to the orbital moment on those electrons; and $B_{dipolar}$ is the dipolar field due to the spin of electrons.

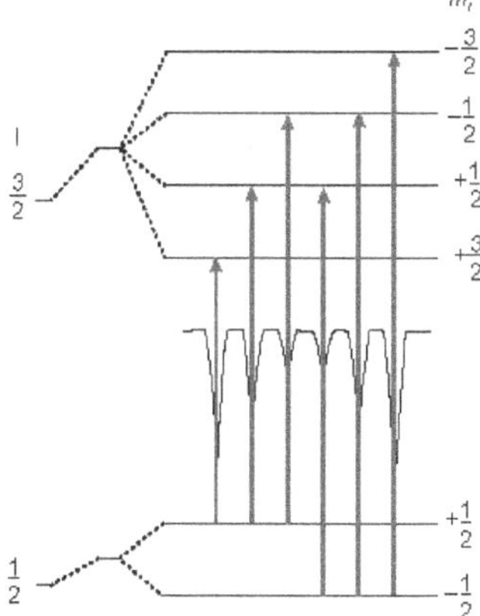

Figure 8.11   Magnetic splitting of nuclear energy levels

The magnetic field splits the nuclear levels with a spin of I into (2I+1) substates as shown in Figure 8.11 for ^{57}Fe. The transitions between the excited state and ground state can only occur where m_I changes by 0 or 1. This gives six possible transitions for a 3/2 to 1/2 transition, giving a sextet as illustrated in Figure 8.11, with the line spacing being proportional to B_{eff}.

The line positions are related to the splitting of energy levels, but the line **intensities** are related to the angle between the Mössbauer gamma ray and the nuclear spin moment. The outer, middle and inner line intensities are given by the relation

$$3:(4\sin^2\theta)/(1+\cos 2\theta):1$$

which means that the outer and inner lines are always in the same proportion but the middle lines may vary in relative intensity between 0 and 4, depending on the angle the nuclear spin moments make to the gamma ray. In polycrystalline samples with no applied field, this value averages to 2 (as shown in Figure 8.11), but in single crystals or under applied fields, the relative line intensities can give information about moment orientation and magnetic ordering.

Applications

Isomer shift, quadrupole splitting and magnetic splitting, singly or in combinations, are the primary characteristics of many Mössbauer spectra. In the following, we show some recorded spectra, which illustrate how measuring these hyperfine interactions can provide valuable information about a system.

Tin Dioxide-Assisted Antimony Oxidation

Antimony-containing tin dioxide is an important catalyst for selective oxidation of olefins. Of particular importance in studying these systems is to know the relative concentrations of antimony charge states (3+ and 5+) during the catalytic process.

Figure 8.12 shows three ^{121}Sb spectra taken at various stages during the catalytic process: (1) fresh Sb_2O_3, (2) Sb_2O_3 calcined at 1000°C and (3) calcined material after catalysis. First, the isotope specificity of Mössbauer spectroscopy picks out only the antimony atoms from the Sn–Sb–O composite. It is apparent from spectrum 1 that practically all of the antimony is in a single state (the red component). A comparison with previous experiments shows that the isomer shift for this majority component matches that of Sb^{3+}. The asymmetric shape is due to quadrupole splitting in this isotope, which has 8 lines (it is a 7/2 to 5/2 transiton).

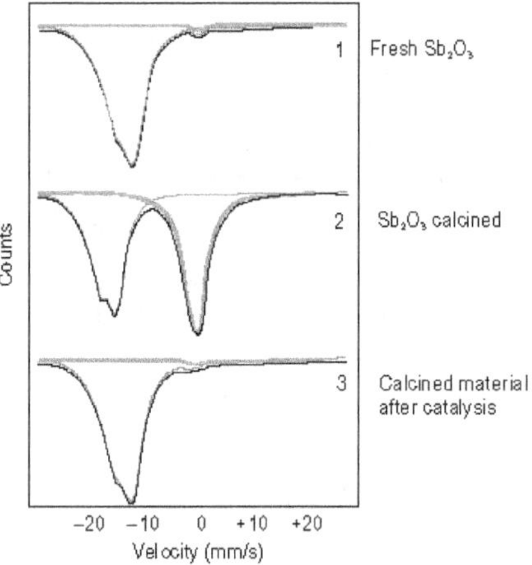

Figure 8.12   Three ^{121}Sb spectra taken at various stages during the catalytic process.

After calcining, the spectrum is now composed of two components of equal area. The second (green) component corresponds to the Sb^{5+} ion. The component areas give the relative proportion of each site within the compound, in this case 1:1 indicating either Sb_2O_4 or Sb_6O_{13}. After catalysis in spectrum 3, we can see that antimony is now in the 3+ charge state again.

Tin spectra were also recorded, showing a single line spectrum of identical isomer shift during all parts of the process, indicating no change in the tin charge state.

In the cases like this, basic deductions can be made even without computer analysis; one can simply see a component appear and disappear and the differences in isomer shift are readily apparent.

Off-centre Tin Atoms in PbSnTeSe

Off-centre impurities are those that can be displaced from their regular positions in a crystal lattice. They can be considered as existing in an asymmetric double potential well. Such atoms can change their position as the temperature changes. Unfortunately, there are often many other phenomena in such systems that mask the off-centring effect.

Mössbauer spectroscopy provides a good tool for observing this effect. Firstly, the movement of the off-centre atom within the lattice will change the symmetry of the electric field, hence changing the quadrupole splitting. Mössbauer spectroscopy is also isotope- and site-specific, which means that we can observe the off-centre single component without any masking from other elements or effects.

A compound that was thought to exhibit off-centring is $Pb_{0.8}Sn_{0.2}Te_{0.8}Se_{0.2}$, with tin as an off-centre atom. The spectra from this sample at 200 K and 20 K are shown in Figure 8.13. There are two components: one from an off-centre site and other from a normal single-potential site. It can be seen in the highlighted region that the small green component develops from a single line to a (broad) doublet. The quadrupole splitting is increasing, indicating that the electric field environment around these particular atoms is becoming more asymmetrical. This is consistent with an atom moving within an asymmetric potential well.

The other component shows no variation in quadrupole splitting. A series of spectra were taken in a temperature cycle and a hysteresis was observed in the values of quadrupole splitting. These results show that tin is an off-centre atom in this compound and that there are two tin sites within it: one normal and one off-centre.

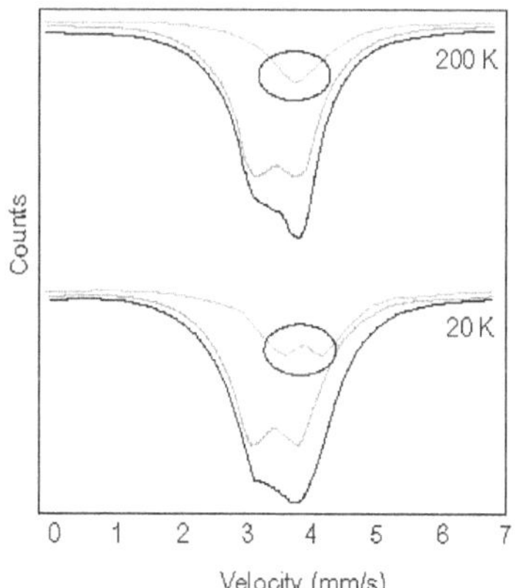

Figure 8.13 ^{119}Sn Mössbauer spectra showing quadrupole splitting as an off-centre atom changes position with a change in temperature.

Uranium/Iron Multilayers

Magnetic multilayers are very important in the areas of data storage and retrieval. A recent development is the use of actinides such as uranium. Uranium in the right environment displays very large orbital magnetic moments, crucial to engineering systems with strong magnetic anisotropy and for magneto-optical applications.

As these samples are sputtered onto a thick substrate, we cannot use conventional Mössbauer spectroscopy in transmission mode (TM) as the substrate would block the gamma rays and we would receive no signal at all. A new technique called conversion electron Mössbauer spectroscopy (CEMS) records the conversion electrons emitted by the resonantly excited nuclei in the absorber. In the TM mode, we record the absorption peaks as the gamma rays are resonantly absorbed and so we see dips, whilst in CEMS we record the electrons emitted from those excited nuclei and so we see emission peaks. As the electrons are strongly attenuated by the sample as they pass through it, most of the signal comes only from the uppermost 1000 Å.

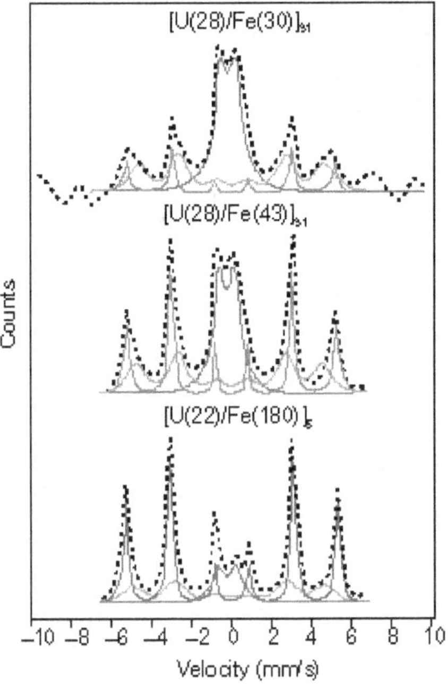

Figure 8.14 ^{57}Fe CEMS spectra taken from U/Fe multilayers of varying layer

Figure 8.14 shows ^{57}Fe CEMS spectra from three uranium iron multilayers of varying layer thicknesses. They are composed of three components: two sextets and one doublet. The hyperfine parameters of the two sextets correspond to alpha iron, the red component being fully crystalline and the green component being from diffuse and poorly crystalline alpha iron. Magnetic splitting shows that iron in these two components is magnetically ordered.

The third component has an isomer shift and quadrupole splitting on the UFe_2 intermetallic. This is paramagnetic at room temperature, as shown by the doublet. It is a doublet and not a sextet even though the intermetallic has a magnetic moment as the moment direction changes much faster in the paramagnetic state than the time it takes for Mössbauer spectroscopy to record it, and thus the experienced hyperfine field averages to zero.

As the iron layer thickness is increased to 43 Å, the relative proportion of alpha iron to UFe_2 increases and also the proportion of fully crystalline iron increases. As the iron layer is further increased to 180 Å, this proportion becomes even greater. We can deduce from this that the thicker the iron layer, the greater the **proportion** of crystalline iron, but a more detailed analysis of component areas as compared to the layer thickness shows that the absolute thickness of the poorly crystalline iron and UFe_2 stay roughly constant.

Mössbauer spectroscopy has easily shown the existence of three different iron sites within the sample and how their proportion has varied with layer thickness.

Superspin Glass Transition in $Al_{49}Fe_{30}Cu_{21}$

The magnetic properties of granular alloys and heterogeneous nanostructures built by ferromagnetic and non-magnetic components attract much attention due to the fundamental interest of their rich phenomenology and to their potential applications, for instance in magneto-resistive devices and magnetic recording. Superspin glasses are of particular interest, but their study is made difficult by the different possible sources for non-equilibrium magnetic behaviour and the mixtures of particle phases within the samples.

Mössbauer spectroscopy, as seen in the previous examples, is very suitable for distinguishing particle sites or phases within a sample. In addition, it can show the difference between magnetically ordered and paramagnetic sites. As the superspin glass phase reaches its freezing temperature, the atoms become magnetically ordered and this will show up in the spectra as a sextet.

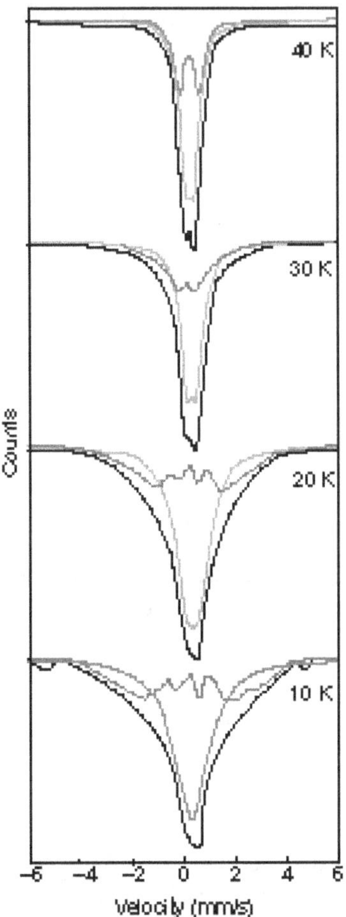

Figure 8.15   Series of ^{57}Fe Mössbauer spectra showing the superspin glass transition in nanogranular $Al_{49}Fe_{30}Cu_{21}$

A series of ^{57}Fe spectra were recorded from a ball-milled sample of $Al_{49}Fe_{30}Cu_{21}$ with decreasing temperature (Figure 8.15). At 40 K, above the freezing temperature, there are two components (i.e., both doublets) of unequal proportion. As the temperature is reduced, the smaller component starts to spread outwards into a magnetic sextet. The peaks are broad and diffuse due to the distribution of grain sizes within the sample and hence the distribution of magnetic hyperfine fields. Plotting the recorded hyperfine fields against temperature can give the superspin glass transition temperature for this compound.

Mössbauer spectroscopy quite readily showed the onset of the superspin glass 'freezing' and the proportion of the magnetic particles and their surrounding non-magnetic matrix. The analysis of hyperfine field distribution also proved consistent with that expected for a superspin glass. The ESR spectra of Fe_3O_4 and Sb_2O_3 are given in Figure 8.16 and Figure 8.17.

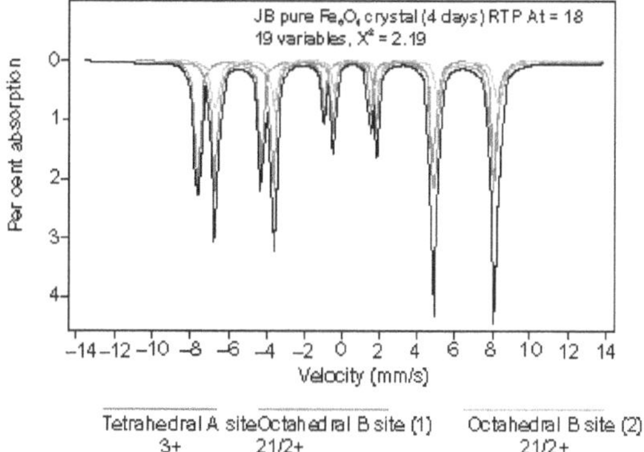

Figure 8.16   Spectrum of magnetite (Fe_3O_4).

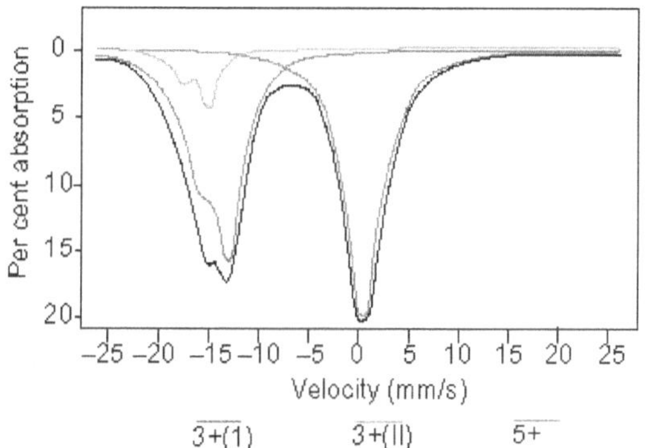

Figure 8.17   A sample of Sb_2O_3 showing both 3+ and 5+ charge states

Structural Deduction of $I_2Br_2Cl_4$

The compound $I_2Br_2Cl_4$ was prepared by oxidizing NaI with $KBrO_3$ followed by chlorine gas. Its structure is likely to be related to that of I_2Cl_6 made from NaI and Cl_2 without treatment by $KBrO_3$. I_2Cl_6 is known to be a planar molecule with a symmetrical Cl-bridged structure (Figure 8.18a). There are several structures for $I_2Br_2Cl_4$ even within this geometric framework. The two Br atoms could be both bridging (Figure 8.18b) or both terminal but on different I atoms (Figure 8.18c and 8.18d) or both terminal but on the same I atom (Figure 8.18e) or one bridging and one terminal (Figure 8.18f).

This question is answered by Mössbauer spectroscopy. The parameters from the ^{129}I Mössbauer spectrum are given in Table 8.1.

Table 8.1 Mössbauer parameters for I_2Cl_6 and $I_2Br_2Cl_4$

Compound	Δ/mm s^{-1}	e^2qQ/MHz
I_2Cl_6	3.50 ± 0.10	$+3060 \pm 10$
$I_2Br_2Cl_4$	I_A 2.82 ± 0.02	$+2916 \pm 10$
	I_B 3.48 ± 0.02	$+3040 \pm 10$

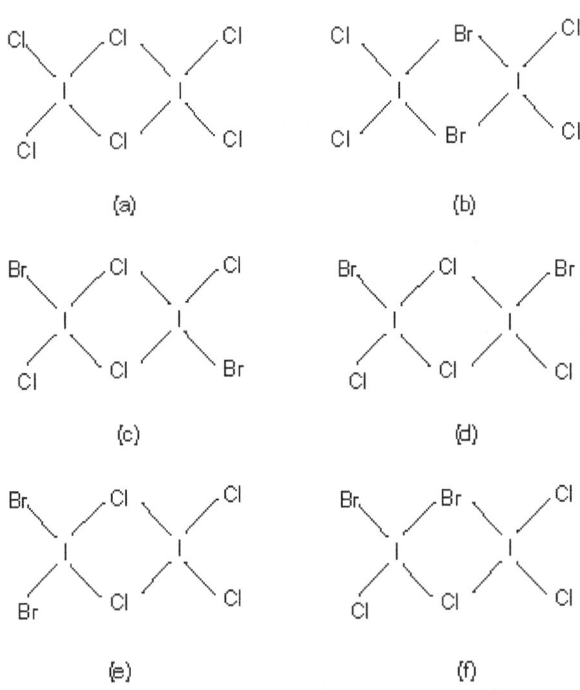

Figure 8.18 Spectrum of I_2Cl_6

In the spectrum of I_2Cl_6, there was one set of line, but in the spectrum of $I_2Br_2Cl_4$, there were two sets of lines. This is due to non-equivalent iodine atoms. The sets were of equal intensity. If the molecule has a single structure, the two iodine atoms cannot be equivalent and so the structures 8.18b and c are ruled out. The isomer shift and quadrupole coupling constant one of iodine atoms are close to those found for I_2Cl_6. This implies that one of iodine atoms in $I_2Br_2Cl_4$ is in much the same environment as are the two in I_2Cl_6, that is, bound to two terminal and two bridging Cl atoms. In structure 8.18d one of the iodine atoms satisfies this condition, but in structure 8.18f neither does; both qh are bound to at least one bromine atom. Hence, the correct structure must be 8.18 e.

Spectrum of Fe(CO)$_5$

For Fe ^{57} the ground state has an I value of 1/2, but the first excited state has a value of 3/2. When there is no field gradient, the excited levels are degenerated. In the presence of field gradient, these levels are split. In axially symmetric fields (linear, square planar), there are I + 1/2 different levels for half integral spins. There are I + 1 different levels for integral spins. The splitting of excited state will not occur in a spherically symmetrical field. A field gradient exists in a trigonal bipyramid molecule Fe(CO)$_5$. Hence, the excited state is split. The two expected transitions give rise to a doublet in the spectrum (Figure 8.19).

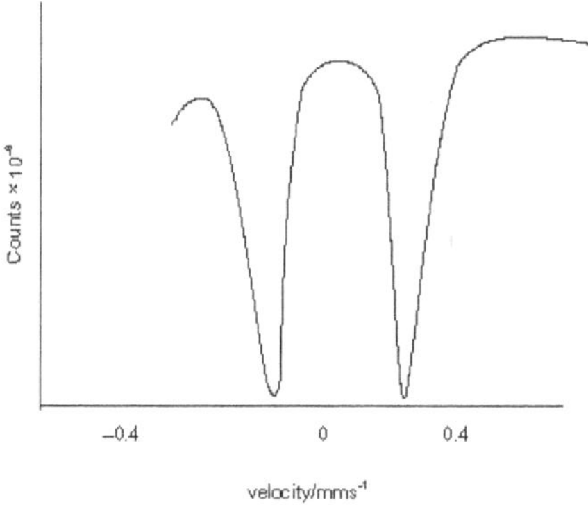

Figure 8.19   Mössbauer spectrum of Fe(CO)$_5$

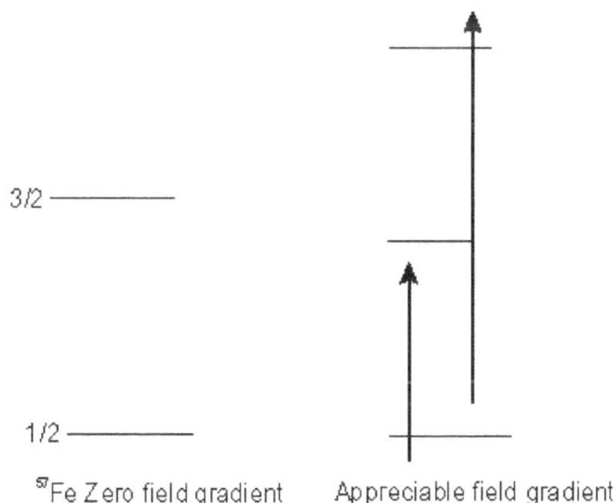

Spectrum of Fe[Fe(CN)$_6$]

The electrostatic interaction between the nucleus and its surrounding electrons can shift nuclear energy levels. Consider the Mössbauer spectrum of Fe[Fe(CN)$_6$], i.e., Fe^{3+}FeIII (CN)$_6$. FeIII is a strong-field iron and Fe^{3+} is a weak-field iron. This substance contains iron in two different chemical environments. Thus, two different energy gamma rays are required to effect transitions.

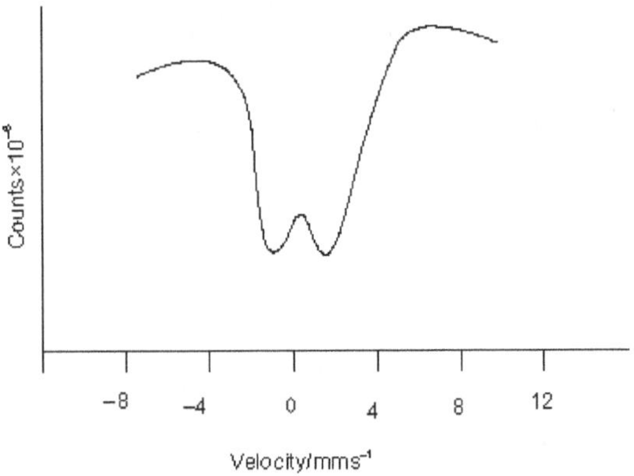

Figure 8.20   Mössbauer spectrum of Fe [Fe (CN)$_6$]

Two peaks are found in the Mössbauer spectrum of this compound (Figure 8.20). One peak at 0.003 cm s^{-1} is assigned to Fe(III) and that at 0.053 to Fe^{3+} cation.

REVIEW QUESTIONS

1. What is Mössbauer spectroscopy? Discuss its applications.

2. Give the conditions under which Mössbauer effect is most likely to occur.

3. What is isomer shift? Explain isomer shift for iron and tin compounds.

4. What are the factors on which isomer shift depends?

5. How are Mössbauer nuclides formed?

6. Mention three nuclides used in Mössbauer spectroscopy.

7. Define recoil-free emission.

8. Explain quadrupole splitting using the Mössbauer spectrum of an iron compound.

9. Discuss Zeeman splitting with an example.

Chapter 9

ATOMIC SPECTROMETRY

Introduction

Absorption and luminescence spectrometry is one of the most broadly used analytical methods in chemistry and biochemistry, including clinical analysis. The term atomic spectrum means the spectrum of atoms or ions. Either may be used to determine the elemental content of analytes.

In the case of free atom, the absorption or emission of gas phase atoms or ions in the UV or visible range is measured.

If the solution of a metallic salt is aspirated into a flame, a vapour containing the atoms of metal may be formed. Some of these gaseous metal atoms may be raised to an energy level that is high enough to permit the emission of radiation imparted by sodium. This is the basis of flame emission spectroscopy (FES) or flame photometry.

However, a large number of atoms will remain in the ground state. These atoms are capable of absorbing radiant energy of their own specific resonance wavelength. In general, it is the wavelength of radiation the atoms would emit when excited from the ground state. Hence, if light of resonance wavelength is passed through a flame containing atoms, then a part of light will be absorbed. The extent of absorption will be proportional to the number of ground state atoms present in the flame. This is the principle of atomic absorption spectroscopy (AAS).

Atomic fluorescence spectroscopy (AFS) is based on the re-emission of absorbed energy by the free atoms.

Instrumentation

FES requires (a) a sample introduction system, (b) a source of atomic vapour (flame), (c) a wavelength isolation system, (d) a radiation detector and (e) a read-out system.

In AAS, an additional subsystem is required, i.e., a primary source of light or radiation. Light from this primary source is passed through the atomic vapour and the amount of radiation absorbed by an atomic species is measured.

As with AAS, AFS also requires a primary source of radiation. The purpose of primary source, i.e., in a 90° configuration, is to generate atomic fluorescence signal from the atomic vapour (Figure 9.1 and 9.2)

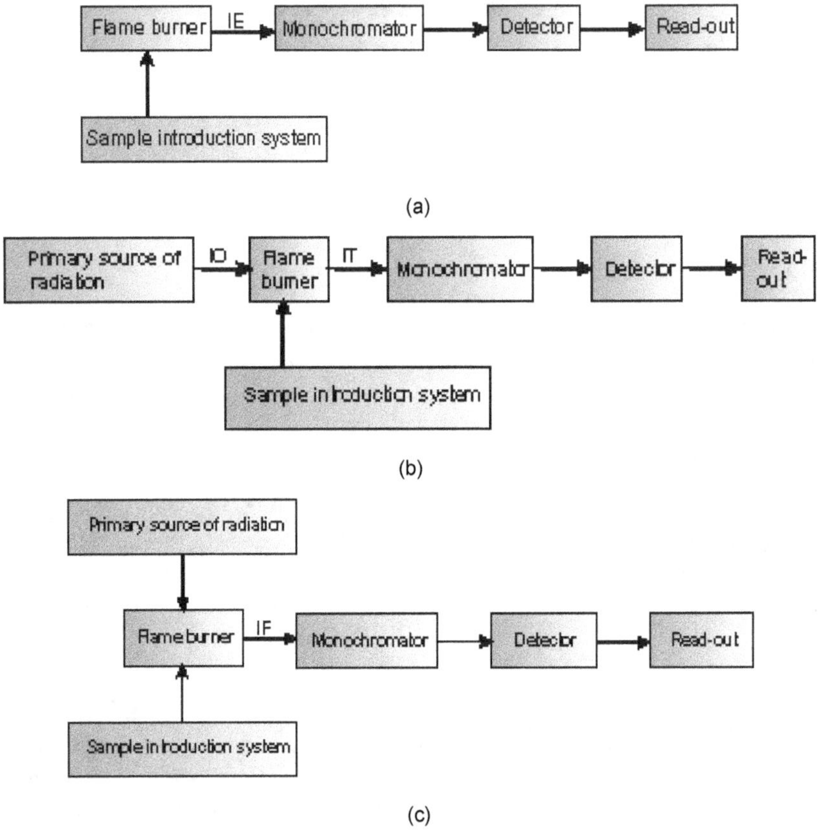

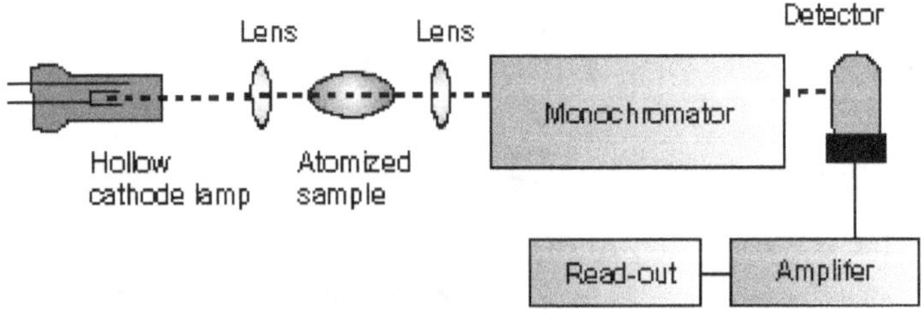

Figure 9.1 appears above with three block diagrams (a), (b), (c):

(a)

(b)

(c)

Figure 9.1   Block diagrams of instrumentation (a) FES (b) AAS (c) AFS

Figure 9.2   AAS instrumentation using hollow cathode lamp

Light Source

The light source is usually a hollow cathode lamp of the element that is being measured. Lasers are also used in research instruments. Since lasers are intense enough to excite atoms to higher energy levels, they allow AA and atomic fluorescence measurements in a single instrument. The disadvantage of these narrow-band light sources is that only one element is measurable at a time.

Hollow Cathode Tube

As with any absorption measurement, a source of radiation that may be absorbed by the sample is required in AAS. For molecular spectrometry, this can be easily achieved by using the combination of a broadband source and a monochromator. The band of radiation emergent from the exit slit of a good monochromator is sufficiently narrow (1 nm) to enable one to profile a UV absorption band whose width is 10 nm.

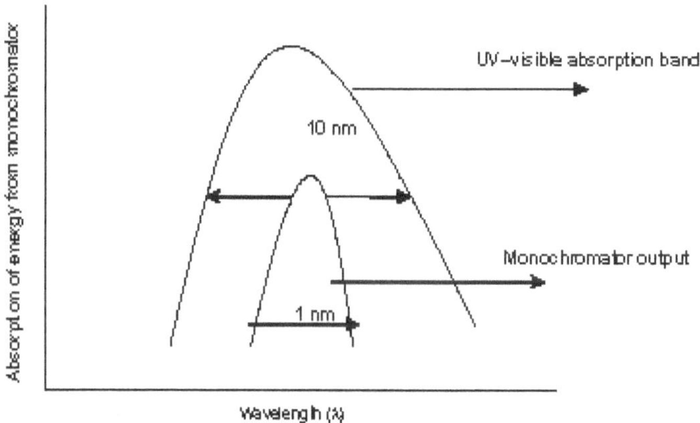

Figure 9.3   UV–visible absorption band and monochromator output

However, the atomic absorption line is much narrower (0.001 to 0.005 nm), and thus absorption profile cannot be easily measured. Most of the radiation from a monochromator exit slit lies outside the AA line. Thus, it cannot be absorbed. Even if such a bandpass is achieved using an advanced monochromator, the resulting output is so low that little light is left for absorption measurement. Hence, an element-specific line source like hollow cathode lamp is required (Figure 9.3) for measurement.

The tube contains an anode and a hollow cylindrical cathode in an inert gas atmosphere at low pressure. The cathode of the lamp is normally constructed from a single element and the spectrum emitted by the lamp is the spectrum of the element and the filler gas. The tube is operated with a power supply of 300 V. Atoms of the gas are ionized in the electrical discharge and energetic Ne or Ar ions are accelerated to the cathode. The collision with the surface dislodges atoms of the cathode metal. This is called *sputtering*. With further collisions, sputtered atoms are excited. In the cooler regions of the discharge, they emit a line spectrum characteristic of the cathode metal (Figure 9.4 and 9.5). This appears as a glow within the cavity of the hollow cathode. A resonance line is selected from this spectrum for absorption measurement. The line width of atomic emission is about the same as the atomic absorption profile of free atoms in the flame.

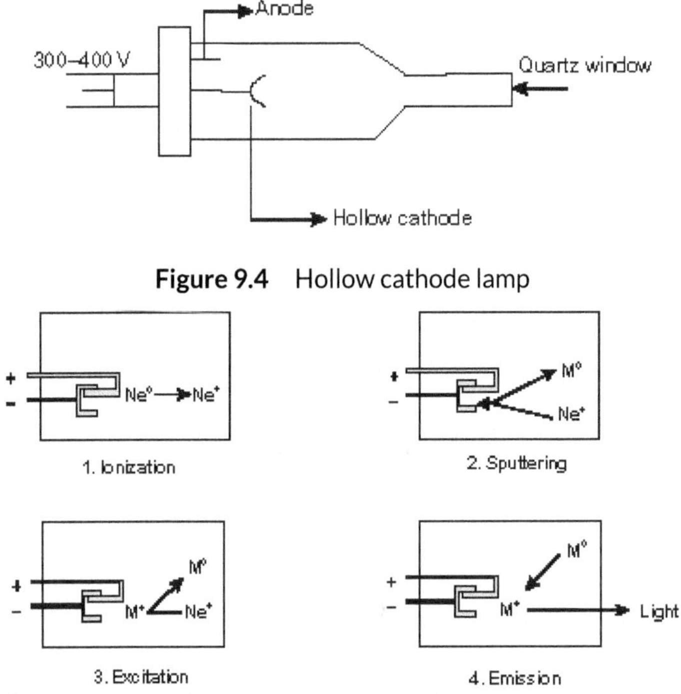

Figure 9.4 Hollow cathode lamp

Figure 9.5 Processes in hollow cathode lamp

Atomizer

AA spectroscopy requires the analyte atoms being in the gas phase. Ions or atoms in a sample must undergo desolvation and vaporization in a high-temperature source such as a flame or graphite furnace. Flame AA can only analyse solutions, while graphite furnace AA can accept solutions, slurries or solid samples.

The sample solutions are usually aspirated with the gas flowing into a nebulizing/mixing chamber to form small droplets before entering the flame.

Flames

Two types of flames are widely used: air–acetylene and nitrous oxide–acetylene. With these flames (Table 9.1), most analytes can be determined.

Table 9.1 Maximum flame temperatures

Oxidant	Fuel	Maximum temperature (°C)
Air	Acetylene	2250
Nitrous oxide	Acetylene	2955
Air	Coal gas	1825
Air	Propane	1725
Air	Hydrogen	2045
Oxygen	Hydrogen	2677
Oxygen	Cyanogens	4500

For most elements, air–acetylene is suited. Its lower temperature favours the formation of neutral atoms. With fuel-rich flame, oxide formation can be minimized. However, if the element is refractory, higher-temperature nitrous oxide–acetylene flame is recommended. The flame has a zone containing high concentrations of CN and NH species, which results in a strongly reducing region, inhibiting the formation of refractory oxides.

The spectral features of air–acetylene and nitrous oxide–acetylene flames show extensive emission from species such as C_2 (500 nm), CH (430 nm) and OH (300 nm). The nitrous oxide–acetylene flame, in addition, has a strong emission from CN (350, 380 nm). The emission intensities are dependent on fuel–oxidant ratio, and fuel-rich conditions tend to increase them. For most AAS measurements, emission from these species is not very important. But for atomic emission measurements, care must be taken to avoid spectral overlap of these complex emissions with analyte emission lines.

Furnaces

Furnaces are used primarily to atomize solids, slurries and solutions for AAS.

Flame atomizers can be used for analysis of liquids or solutions by AAS.

Flame AA uses a slot-type burner (Figure 9.6) to increase the path length, and therefore to increase total absorbance.

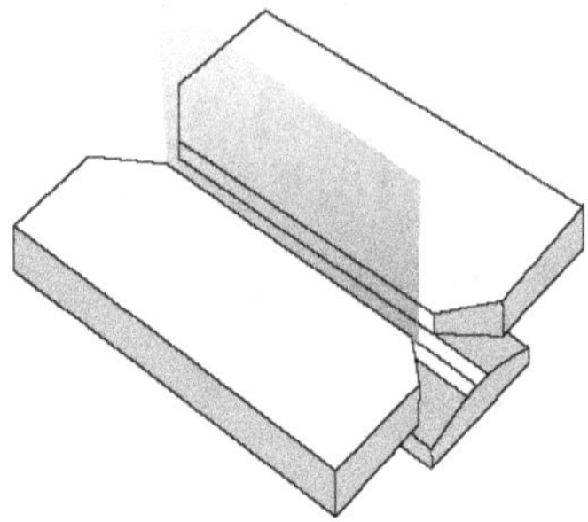

Figure 9.6   Slot-type burner

Flame sources of atomization have been brought to a high degree of reliability. However, since the gases for the flame dilute the sample, a long flame shape is desired to obtain high sensitivities. The flames are replaced by inert gas plasmas. Plasmas provide overall better emission results for a wide range of elements than do flames and are, therefore, better suited for simultaneous multi-element analysis. Gas plasmas do not require tanks of flammable gases in the laboratories.

Plasma is an ionized gas. High-temperature plasmas function as high-temperature atomizers. The most commonly used type for atomic emission is an inductively coupled plasma (ICP).

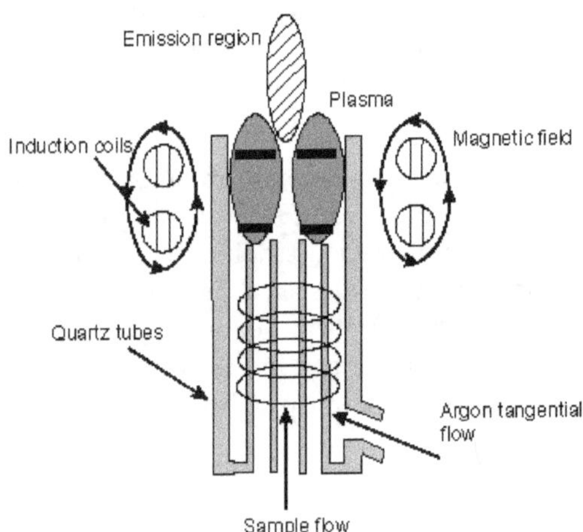

Figure 9.7   Inductively coupled plasma

ICP (Figure 9.7) is a continuous plasma induced in a flowing stream of argon. The energy to ionize argon is added through a radio frequency electromagnetic field. The electrons are accelerated by the field and hit the surrounding atom causing heating and further ionization, which sustains the plasma state. The named inductively coupled plasma arises because plasma is not directly connected to the loop. The energy is added inductively. If the gas flow and radio frequency power are regulated properly, a doughnut-shaped region with a high, constant temperature is sustained at the base of the plume. This is the region into which the sample is introduced.

Other methods used to generate continuous plasmas are: (i) plasmas excited by a direct current voltage passing across a gap (a DC plasma or spark source or arc), (ii) plasmas excited by an alternating current (an AC plasma), (iii) plasmas introduced by centimetre wavelength radiation (a microwave induced plasma) and (iv) glow discharges generated at surfaces of metals. None of these is as common as the ICP.

Electrodeless Discharge Tubes

A small amount of the element or salt of the element is sealed inside a quartz bulb with a small amount of inert gas. This bulb is placed inside a ceramic cylinder on which a helical resonator is coiled. When a radio frequency is applied, the energy ionizes the inert gas and then excites the metal atoms. This gives the emission spectrum (Figure 9.8).

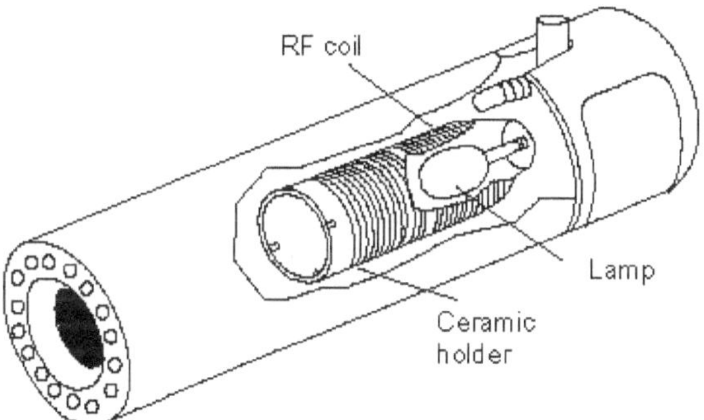

Figure 9.8   Electrodeless discharge tube

Hollow Graphite Tube

A hollow graphite tube with a platform is shown in Figure 9.9. 25 ml of the sample (approximately 1/100th of a raindrop) is placed through the sample hole and on to the platform from an automated micropipette and sample changer. The tube is heated electrically by passing a current through it in a pre-programmed series of steps. The details will vary with the sample, but typically they might be 30–40 seconds at 150°C to evaporate the solvent, 30 seconds at 600°C to drive off any volatile organic material and char the sample to ash, and with a very fast heating rate (approximately 1500°C s^{-1}) to 2000–2500°C for 5–10 seconds to vaporize and atomize elements (including the element being analysed). Finally, heating the tube to a still higher temperature approximately 2700°C, cleans it ready for the next sample. During this heating cycle, the graphite tube is flushed with argon gas to prevent the tube burning away. In electrothermal atomization, almost 100% of the sample is atomized. This makes the technique much more sensitive than flame AAS.

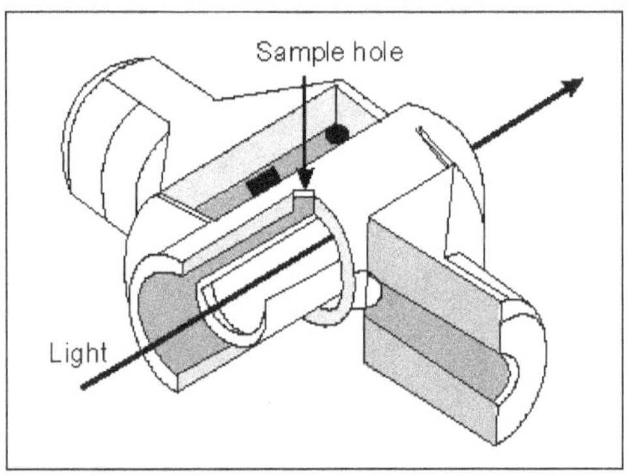

Figure 9.9 Hollow graphite tube

Monochromators and Detectors

AA spectrometers use monochromators and detectors for UV and visible light. The main purpose of the monochromator is to isolate the absorption line from background light due to interferences. Simple dedicated AA instruments often replace the monochromator with a bandpass interference filter. Photomultiplier tubes are the most common detectors for AA spectroscopy.

Atom Formation—Steps and Processes

The main steps involved in atom formation are:

i. ***Solution transport***   The transport involves the movement of solution to the nebulizer through a small bore plastic tube. An identical and reproducible uptake rate is required for samples and standards. Solution viscosity and surface tension changes, between samples and standards must be avoided. Motor-driven peristaltic pumps or syringe drives can be used to ensure accurate and precise delivery of the sample solution to the nebulizer.

ii. ***Nebulization***   In this process, the solution is converted into a fine spray or aerosol. This is accomplished with a concentric pneumatic nebulizer. Typically, the droplet diameters should be less than 10 μm and preferably approaching 1–2 μm. The net result is that about 90% of the sample goes down the drain at this step.

iii. ***Aerosol transport***   The third step is transporting the aerosol to the flame through the spray chamber. The spray chamber and its subcomponents like impact beads, impactors, and the geometry of the spray chamber ensure that only the finest aerosol actually reaches the flame. Large droplets are condensed and drained off.

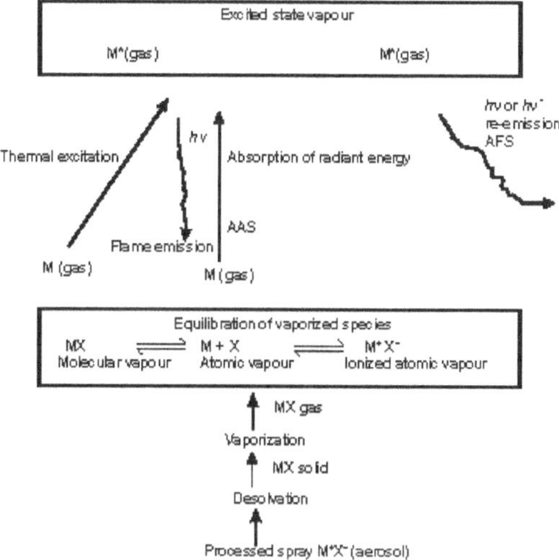

iv. ***Desolvation***   Once the aerosol reaches the flame, it is converted into salt particles. The rate of desolvation depends on droplet size, time i n flame, velocity of the droplet and flame temperature. As long as only fine aerosol is allowed to reach the flame, incomplete desolvation is seldom a problem.

v. ***Vaporization***   It is the critical step in flame photometric method. This step is defined as the conversion of a salt particle into vapour in the flame. In this step, interferences occur.

vi. ***Equilibration of vaporized state***   This equilibrium is between molecular species (often monoxides and monohydroxides) and neutral atom and ions (normally singly ionized). In the flame method, free neutral atoms are desired. In order to minimize the formation of molecular species, reducing flame conditions are required.

Factors Affecting Atomization/Ionization

Ideally in AAS technique, all atoms are freed of the influence of the surrounding matrix. This means that the analyte atoms are not chemically bound to any other atom. All particles, compounds and molecules must be dissociated. **Atomized** is the correct term. In addition, the matrix of the sample should be destroyed totally.

Solid Samples

Atomization is a complex process that is still not well understood in detail. The order of major steps in the process has been elucidated. Atomization of a powdered solid material by rapid heating proceeds in general through the following steps.

1. The powder heats up.

2. Particles that are composed of many atoms break off from the powder surface.

3. The multi-atom clusters vaporize and break up (atomize) into atoms and ions.

If these three steps do not proceed to completion and produce atoms that remain detectable for a long enough time, then achieving good precision or accuracy may be compromised. The problems may occur in two ways: (1) Refractory (thermally stable) compounds may be formed during the heating step. (2) Volatile compounds may form, and they escape before they can be detected. In the first case, refractory oxides are more common but stable nitrides or carbides can also be formed. Metals may interact with chlorine atoms and form volatile metal chlorides. In either instance, a part of the total elemental content could go undetected.

The errors associated with incomplete atomization and molecule formation can be generally minimized by using higher temperatures. However, elevating the temperature can shift the gas phase equilibrium from atoms towards ionic species, which may not be desirable.

Liquid Samples

When the sample to be atomized is a liquid, a few additional steps are involved. The samples are introduced into the flame as aerosols. The aerosol is generated using pneumatic or ultrasonic-shear nebulizer.

In a pneumatic nebulizer, the sample solution is drawn in by rapid flow of oxidant and it is turned into a fine mist when it leaves the tip of the nebulizer and strikes the glass bead. The formation of small droplets is called nebulization. The droplets range from the sizes of fog to fine raindrops. This heterogeneous mixture interacts with a baffle or skimmer in the gas stream that breaks up or removes the larger droplets. The larger droplets move sluggishly, strike a surface and either break up or flow out as drain. As a result, 10% of the liquid reaches the thermal atomizer (Figure 9.10).

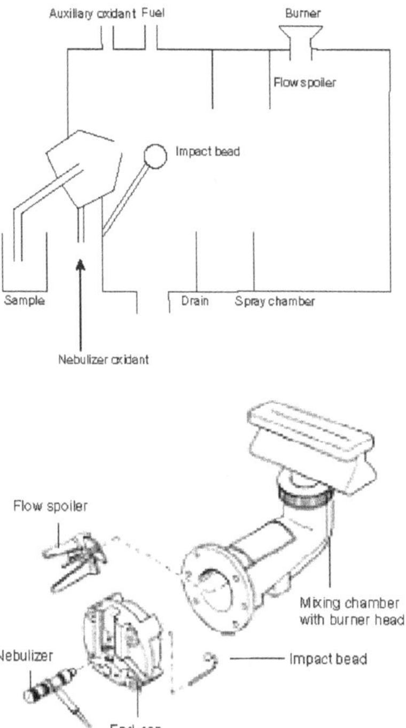

Figure 9.10 Parts of nebulizer

Ultrasonic vibration of a capillary can shake free drops to form a mist. These are the bases of ultrasonic nebulizers. Ultrasonic nebulizers have decreased the limits of detection by two orders of magnitude relative to the pneumatic nebulizers.

Nomenclature of Atomic Spectroscopy

Sensitivity In the past, the term sensitivity in atomic absorption indicated the concentration or mass of analyte needed to produce a signal with a transmittance of 0.99 or its equivalent absorbance of 0.0044. Nowadays, the amount needed to produce an absorbance of 0.0044 is called 'characteristic concentration' or 'characteristic mass'. In other words, the term sensitivity refers to the slope of working curve in the analytical concentration range.

Detection limit It is defined as the amount of concentration of analyte needed to produce a signal three times the peak-to-peak noise level of the blank.

Wavelength is measured in nm.

Spectral notation The atoms from which atomic spectra arise can exist in a number of different formal oxidation states. For example, copper could be atomic or ionic: Cu^0, Cu^+ Cu^{2+}. The nomenclature of atomic spectra differs from that used to describe the chemistry of solutions and solids. The spectroscopic nomenclature uses roman numerals that are one unit higher than the oxidation states. The neutral atom is I, the oxidation state +1 is II, and the +2 oxidation state is III.

In atomic spectroscopy	Cu(I)	Cu(II)	Cu(III)
In inorganic chemistry	Cu(0)	Cu(I)	Cu(II)
	Cu^0	Cu^+	Cu^{2+}

Atomic absorption spectrum is obtained by plotting intensity versus wavelength or wave number.

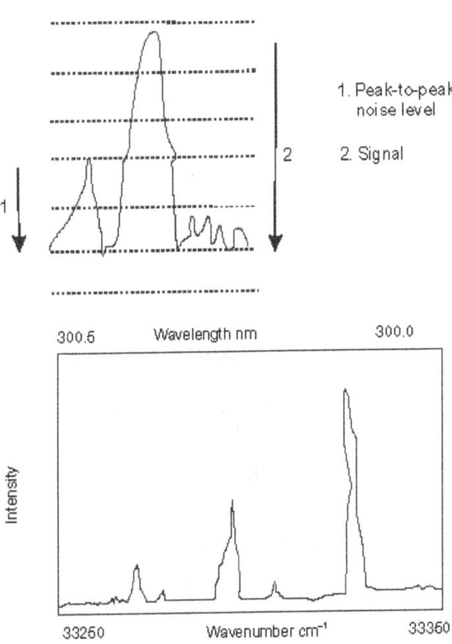

Problem

A 12-ppm solution of lead gives an atomic absorption signal of 30% absorption. What is the atomic absorption sensitivity?

$$70\%T = 0.7\,T \quad A = \log 1/T = \log 1/0.7 = 0.1550$$

$$Sensitivity = \frac{12 \times 0.0044}{0.1550} = 0.34\,ppm$$

Interferences

Interference is any effect that changes the signal when analyte concentration remains unchanged. Interference can be corrected by countering the source of interference or by preparing standards that exhibit the same interference.

The factors that interfere with the determination of concentration of an element are classified as (a) spectral and (b) chemical interferences.

Spectral Interference

Spectral interference or atomic line overlap in AAS arises mainly from the overlap between the frequencies of a selected resonance line with lines emitted by some other element. For example, europium (λ 324.755 nm) at a concentration of 150 ppm would interfere with copper (324.754 nm). Similarly, mercury (253.652 nm) would be masked by cobalt at concentrations above 200 ppm (253.649 nm).

In FES, when the line emission of the element to be determined and those due to interfering substances are of similar wavelength, spectral interferences occur.

These interferences can be eliminated by improved resolution of the instrument, e.g. by the use of a prism rather than a filter. Phosphate can interfere with atomic absorption spectrum of calcium. This is eliminated by allowing more time for vaporization, i.e., by increasing the height of the flame. The interference of aluminium in the determination of magnesium is eliminated by using N_2O–C_2H_2 flame. Higher temperature is effective in vaporizing refractory salt particles. For example, Al_2O_3 is more easily dissociated in N_2O–C_2H_2 flame.

Chemical Interferences

These interferences occur as a result of stable compound formation and ionization.

Stable compound formation leads to incomplete dissociation of the substance to be analysed when placed in a flame. For example, in the determination of calcium in the presence of phosphate or sulphate and formation of stable refractory oxides of titanium, vanadium and aluminium can be overcome by the following ways:

i. By the use of releasing agents. Consider the following reaction.

$$M—X + R \leftrightarrow R—X + M$$

An excess of releasing agent R will lead to an enhanced concentration of the required gaseous atom M. Thus, in the determination of calcium in the presence of phosphate, addition of excess lanthanum chloride or strontium chloride will give rise to lanthanide or strontium phosphate. Calcium can be determined in acetylene–air flame without any interference.

ii. Extraction of the analyte or interfering element can be carried out prior to analysis. The common methods include solvent extraction and ion exchange.

iii. Increase in flame temperature leads to free gaseous atoms. For example, aluminium oxide is more readily dissociated in acetylene–nitrous oxide flame than in acetylene–air flame. Calcium–aluminium interferences arising from the formation of calcium aluminate can also be overcome by using acetylene–nitrous oxide flame.

Ionization interference is a problem in the analysis of alkali metals that have the lowest ionization potentials. For any element, the gas-phase ionization reaction can be written as:

$$M(g) \rightleftharpoons M^+(g) + e^-(g) \qquad (1)$$

At 2450 K, sodium is 5% ionized. Because ionized atoms have energy levels different from those of neutral atoms, the desired signal is decreased.

An ionization suppressor is an element added to the sample to decrease the ionization of analyte. For example, 1 mg/mL of CsCl is added to the sample for the analysis of potassium, because caesium is more easily ionized than potassium. By producing a high concentration of electrons in the flame, ionization of Cs reverses the equation (1).

Drawbacks   The primary instrumental disadvantage of AAS is that generally only a single element can be measured at a time. A different light source is required for each element. Thus, multi-element analysis is very tedious and time-consuming. AAS is essentially only a quantitative technique and cannot be effectively used for qualitative analysis.

AAS Methodology

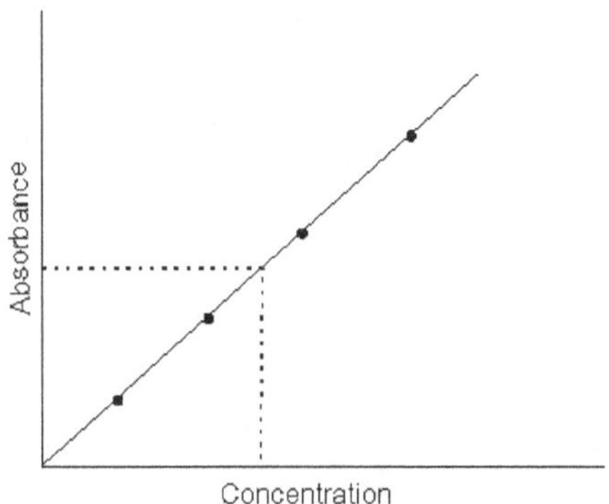

Figure 9.11   Calibration curve

The atomic absorption spectrometer is operated in the absorption mode. For a given element, a series of standard solutions are prepared. The appropriate source is selected and the blank solution is introduced. The absorbance is set at zero for the blank. The standard solutions are introduced one by one and the absorbances are read. The sample solution is then fed into the instrument and the absorbane is read.

Atomic Emission Methodology

A flame emission spectrometer operates much like an atomic absorption spectrometer but without the hollow cathode lamp.

Regardless of how a sample is atomized, the light from atomic emission contains lines characteristic of all or most of the elements in the sample. It is possible, therefore, to determine a number of different elements from the same source. Emission spectroscopy is intrinsically a simultaneous multi-element analysis technique.

With the most sophisticated equipment, concentrations of 70 elements can be determined with a single experiment. The optimum experimental conditions for various elements differ. The elements will exhibit differences in atomization depending on their binding energies in the sample matrix. Some elements atomize quickly and others slowly.

In an ICP torch, the easily atomized elements should show up near the bottom of the torch, where the heating starts, than the more slowly atomized species. For example, the relative intensities of emission from atomic phosphorus and singly ionized copper differ significantly at different heights in the plume. Thus, the conditions optimized for one element are seldom best for all of the other elements present.

Any change in the gas flow rate, radio frequency power and temperature can change the chemical species and their emissivities in the volume that is monitored by the detector. When these changes are random, the chemical changes contribute directly to the noise. As a result, detection limit and precision depend directly on the stability of plasma chemistry.

One of common applications of flame emission spectrometry is the determination of alkali metals, particularly in clinical laboratory. The instrument designed specifically for this determination may contain a single interference filter as the monochromator and a vacuum phototube as the detector.

Atomic Fluorescence Methodology

This technique incorporates aspects of both atomic absorption and atomic emission. Like atomic absorption, ground-state atoms created in a flame are excited by focusing a beam of light into the atomic vapour. Instead of looking at the amount of light absorbed in the process, however, the emission resulting from the decay of atoms excited by the source light is measured. The intensity of "fluorescence" increases with increasing atom concentration, providing the basis for quantitative determination.

The source lamp for atomic fluorescence is mounted at an angle to the rest of the optical system, so that the light detector sees only the fluorescence in the flame and not the light from the lamp itself. It is advantageous to maximize lamp intensity since sensitivity is directly related to the number of excited atoms, which in turn is a function of the intensity of exciting radiation.

The detection limits for select elements are given in the Table 9.2.

Table 9.2   Detection limits (mg/ml) for selectd elements

Element	AAS Flame	AAS Electrothermal	AES Flame	AFS ICP	APS Flame
Al	30	0.005	5	2	5
As	100	0.02	0.0005	40	100
Ca	1	0.02	0.1	0.02	0.001
Cd	1	0.0001	800	2	0.01
Cr	3	0.01	4	0.3	4
Cu	2	0.002	10	0.1	1
Fe	5	0.005	30	0.3	8
Hg	500	0. 1	0.0004	1	20
Mg	0.1	0.00002	5	0.05	1
Mn	2	0.0002	5	0.06	2
Mo	30	0.005	100	0.2	60
Na	2	0.0002	0.1	0.2	–
Ni	5	0.02	20	0.4	3
Pb	10	0.002	100	2	10
Sn	20	0.1	300	30	50
V	20	0.1	10	0.2	70
Zn	2	0.0005	0.0005	2	0.02

Bulk Analysis without Atomization

Bulk elemental techniques do not require atomization in order to obtain elemental analysis. This technique includes X-ray fluorescence and neutron activation analysis.

Both these methods are non-destructive. X-ray fluorescence is discussed in the following chapter. Neutron activation analysis (NAA) is a sensitive analytical technique useful for performing both qualitative and quantitative multi-element analysis of major, minor and trace elements in samples from almost every conceivable field of scientific or technical interest. For many elements and applications, NAA offers sensitivities that are superior to those attainable by other methods, on the order of parts per billion or better. In addition, because of its accuracy and reliability, NAA is generally recognized as the 'referee method' of choice when new procedures are being developed or when other methods yield results that do not agree. It is estimated that approximately 1,00,000 samples undergo NAA each year world over.

Neutron activation analysis was discovered in 1936 when Hevesy and Levi found that samples containing certain rare earth elements became highly radioactive after exposure to a source of neutrons. From this observation, they quickly recognized the potential of employing nuclear reactions on samples followed by measurement of the induced radioactivity to facilitate both qualitative and quantitative identification of elements present in the samples.

The basic essentials required to carry out an analysis of samples by NAA are a source of neutrons, instrumentation suitable for detecting gamma rays, and a detailed knowledge of the reactions that occur when neutrons interact with target nuclei. A brief description of the NAA method, reactor neutron sources and γ-ray detection is given below.

The NAA Method

The sequence of events occurring during the most common type of nuclear reaction used for NAA, namely the neutron capture or (n,γ) reaction, is illustrated in Figure 9.12. When a neutron interacts with the target nucleus via a non-elastic collision, a compound nucleus forms in an excited state. The excitation energy of the compound nucleus is due to the binding energy of the neutron with the nucleus. The compound nucleus will almost instantaneously de-excite into a more stable configuration through the emission of one or more characteristic prompt gamma rays. In many cases, this new configuration yields a radioactive nucleus, which also de-excites (or decays) by the emission of one or more characteristic delayed gamma rays but at a much slower rate according to the unique half-life of the radioactive nucleus. Depending upon the particular radioactive species, half-lives can range from fractions of a second to several years.

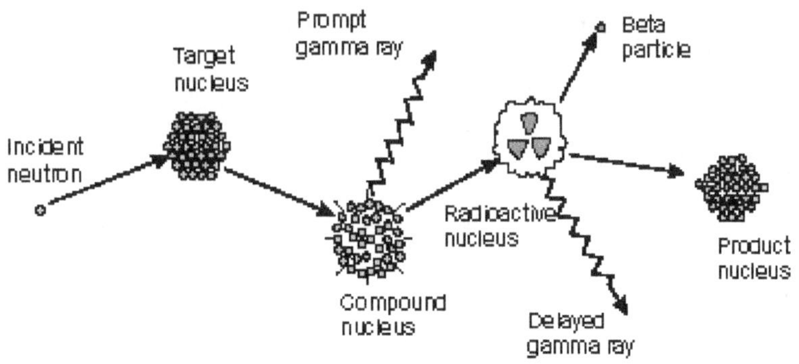

Figure 9.12   Process of neutron capture by a target nucleus followed by the emission of gamma rays

In principle, therefore, with respect to the time of measurement, NAA falls into two categories: (1) prompt gamma-ray neutron activation analysis (PGNAA), where measurements take place during irradiation, or (2) delayed gamma-ray neutron activation analysis (DGNAA), where the measurements follow radioactive decay. The latter operational mode is more common; thus, when one mentions NAA it is generally assumed that the measurement of delayed gamma rays is intended. About 70% of the elements have properties suitable for measurement by NAA.

Neutrons

Although there are several types of neutron sources (reactors, accelerators, and radioisotopic neutron emitters) one can use for NAA, nuclear reactors with their high fluxes of neutrons from uranium fission offer the highest available sensitivities for most elements. Different types of reactors and different positions within a reactor can vary considerably with regard to their neutron energy distributions and fluxes due to the materials used to moderate or reduce the energies of the primary fission neutrons. However, most neutron energy distributions are quite broad and consist of three principal components—thermal, epithermal, and fast.

The thermal neutron component consists of low-energy neutrons (energies below 0.5 eV) in thermal equilibrium with atoms in the reactor's moderator. At room temperature, the energy spectrum of thermal neutrons is best described by the Maxwell–Boltzmann distribution with a mean energy of 0.025 eV and a most probable velocity of 2200 m/s. In most reactors, 90–95% of the neutrons that bombard a sample are thermal neutrons.

Measurement of gamma rays   The instrument used to measure gamma rays from radioactive samples generally consists of a semiconductor detector, associated electronics, and a computer-based multi-channel analyser (MCA/computer). Most NAA labs operate one or more hyperpure or intrinsic germanium (HPGe) detectors which can be operated at liquid nitrogen temperatures (77 degrees K) by mounting the germanium crystal in a vacuum cryostat, thermally connected to a copper rod or "cold finger". Although HPGe detectors come in many different designs and sizes, the most common type of detector is the coaxial detector which in NAA is useful for measurement of gamma rays with energies over the range from about 60 keV to 3.0 MeV.

The two most important performance characteristics requiring consideration when purchasing a new HPGe detector are resolution and efficiency. Other characteristics to consider are peak shape, peak-to-Compton ratio, crystal dimensions or shape, and price.

Determination of Elemental Concentration

The procedure generally used to determine concentration (i.e., ppm of an element) in the unknown sample is to irradiate the unknown sample and a comparative standard containing a known amount of the element of interest together in the reactor. If the unknown sample and the standard are both measured on the same detector, then one needs to correct the difference in decay between the two. One usually corrects the measured counts (or activity) for both samples back to the end of irradiation using the half-life of the measured isotope. The equation used to calculate the mass of an element in the unknown sample relative to the standard is

$$\frac{A_{sam}}{A_{std}} = \frac{m_{sam}}{m_{std}} \frac{(e^{-\lambda T_d})_{sam}}{(e^{-\lambda T_d})_{std}}$$

where,

$A =$ activity of the sample (sam) and standard (std),

$m =$ mass of the element,

$l =$ decay constant for the isotope, and

$T_d =$ decay time

When performing short irradiations, the irradiation, decay and counting times are normally fixed the same for all samples and standards such that the time-dependent factors get cancelled. Thus the above equation is simplified into

$$C_{sam} = C_{std} \frac{W_{std}}{W_{sam}} \frac{A_{sam}}{A_{std}}$$

where,

$C =$ concentration of the element and

$W =$ weight of the sample and standard.

REVIEW QUESTIONS

1. Compare the advantages and disadvantages of furnaces and flames in atomic absorption spectroscopy.

2. What is meant by spectral interference, chemical interference and ionization interference?

3. What is the difference between atomic absorption and atomic emission spectroscopy?

4. Write a note on hollow cathode lamp.

5. Discuss the various steps in atom formation.

6. Describe the basic components of atomic absorption spectrometer.

7. Why is the technique of atomic absorption limited to metals?

8. What is sputtering?

9. List out the applications of AAS.

10. Write briefly on neutron activation analysis.

11. Mention the detectors used in AAS.

12. What is atomic fluorescence?

X-RAY SPECTROMETRY

Introduction

X-rays are electromagnetic waves whose wavelengths range from about 0.1 to 100×10^{-10} m. They are produced when rapidly moving electrons strike a solid target and their kinetic energy is converted into radiation. The wavelength of the emitted radiation depends on the energy of electrons. Since each element has characteristic transition energies, the wavelength of emitted radiation is characteristic of the element being bombarded. X-rays can be used in chemical analysis in three distinct ways.

1. X-ray absorption

2. X-ray diffraction

3. X-ray fluorescence

X-ray absorption analysis utilizes the difference in absorption of X-rays by different materials. Major discontinuity in the absorption of X-rays by an element occur when the energy of the X-rays becomes sufficient to knock an electron out of the inner levels of an atom. With X-ray absorption, it is possible to tell if there is a hole in welded joint on the hull of a ship. It is also possible to measure the volume of a liquid in a storage tank without taking the liquid out of the tank.

A second method of using X-rays is the diffraction of X-rays from the planes of a crystal. This method depends upon the wave character of X-rays and the regular spacing of planes in the crystal. Through X-ray diffraction, one can identify the crystal structures of various solid compounds. The arrangement of molecules in a crystal can be determined.

The X-ray method in common use is X-ray fluorescence, a fluorescence technique carried out with backscatter geometry. If a sample is irradiated with an X-ray beam, the sample sometimes emits other X-ray beams. This process is called X-ray fluorescence. The wavelength of X-ray fluorescence enables one to determine the elements present in a sample. Due to inherent experimental limitations, the elements that may be quantified by X-ray fluorescence are those that are heavier than neon. Also, since almost all elements produce X-rays under the experimental conditions, the 'contrast' is not especially high. Thus, X-ray fluorescence is limited to concentration levels greater than about 0.1%. The method has two advantages.

1. It can be used for the analysis of solids without sample dissolution.

2. It is a multi-element technique.

The fluorescence and diffraction techniques are to a large extent complementary, since one allows accurate quantitation of elements to be made and the other allows qualitative and semiquantitative estimations to be made of the way in which the matrix elements are combined to make up the phases in the specimen. Thus a combination of the two techniques will often allow the accurate determination of the material balance of a sample.

Instrumentation

The components for the three distinct fields of X-ray analysis are the same for each field, but the optical system varies for each field. The main components are X-ray source, collimator, monochromator and detector.

X-Ray Source

The generation of X-rays for use in analysis is done by bombarding a metal slab such as Mo or W with electrons accelerated through a voltage of 30,000 to 50,000 volts. The bombardment occurs in a high vacuum. X-rays thus generated by particle bombardment are called **primary X-rays.**

A typical X-ray generator passes an electric current through a filament, which causes the electrons to be emitted. These electrons are then accelerated by high voltage (usually somewhere between 20 and 100 kV) towards an anode (target). A thin beryllium window is used to allow the X-rays out of the generator, and on to the sample (Figure 10.1).

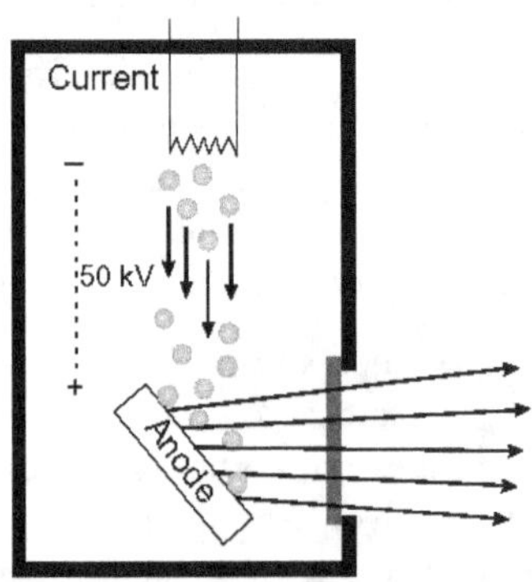

Figure 10.1   Schematic diagram of an X-ray source

The second way to generate X-rays is to illuminate a material with X-rays or γ radiation of a higher energy than those needed for the analysis. X-rays produced upon illumination with X-ray photons are called **secondary X-rays**.

When metals having smaller atomic numbers such as copper or molybdenum are bombarded by electrons, very sharp spikes are observed as shown in Figure 10.2. The broad background emission is called **bremsstrahlung.** This radiation is present only when the X-rays are generated by bombarding with charged particles but not by irradiating with other higher energy X-ray.

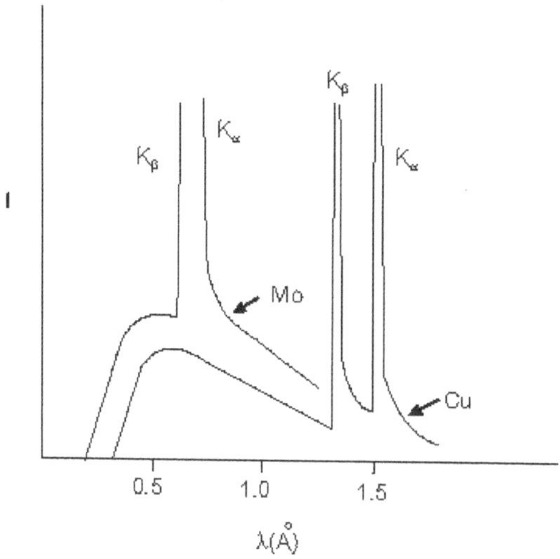

Figure 10.2   Typical X-ray emission

The sharp lines are labelled using the following nomenclature. The highest energy lines, if there are more than one set, are called the K lines. The next highest L and the next M and so on.

The sharp emission lines shown in Figure 10.2 are K lines. The line at higher wavelength (lower energy) is named K$_\alpha$ and the next one at lower wavelength (higher energy) is the K$_\beta$ line. The origin of the K$_\alpha$ and K$_\beta$ are from, the $L \rightarrow M$ $M \rightarrow K$ and transitions in the atom. A unique label for a given line is formed by writing the symbol for the emitting atom first, for example, Mo–K$_\alpha$.

If several metals are present in the target, each will emit its characteristic radiation independently. This property can be used to determine qualitatively the elements that are present in an alloy by making it the target in an X-ray tube and then by scanning all the wavelengths emitted by the target. Typical targets and their related constants are given in Table 10.1.

Table 10.1 Target materials and associated constants

	Cr	Fe	Cu	Mo
Z	24	26	29	42
α_1, Å	2.2896	1.9360	1.5405	0.70926
α_2, Å	2.2935	1.9399	1.5443	0.71354
α^*, Å	2.2909	1.9373	1.5418	0.71069
β_1, Å	2.0848	1.7565	1.3922	0.63225
β, filt.	V, 0.4mil †	Mn, 0.4 mil	Ni, 0.6 mil	Nb, 3 mils
α, filt.	Ti	Cr	Co	Y
Resolution, Å	1.15	0.95	0.75	0.35
Critical potential, kV	5.99	7.11	8.98	20.0
Operating conditions, kV:	30–40	35–45	35–45	50–55
Half- or full-wave-rectified, mA	10	10	20	20
Constant potential, mA	7	7	14	14

$*\overline{\alpha}$ s the intensity-weighted average of α_1 and α_2 and is the figure usually used for the wavelength when the two lines are not resolved.

† 1 mil =0.001 inch = 0.025 mm

Filt: use of appropriate filter

Collimator

The X-rays emitted by the anode are radially directed, and so, they form a hemisphere with the target at the centre. A narrow beam of X-rays can be made by using two sets of closely packed metal plates separated by a small gap. This arrangement absorbs all the radiation except the narrow beam that passes between the gap. Decreasing the distance between the plates or increasing the gap increases resolution but decreases intensity.

Monochromator

Filter A filter is a window of materials that absorbs unwanted radiation but permits radiation of the desired wavelengths to pass. For example, zirconium filter is used for molybdenum. The zirconium strongly absorbs the radiation of molybdenum at shorter wavelengths, but weakly absorbs the molybdenum K_∞ line. The net result is that the radiation passing the zirconium filter is purer and the ratio of the molybdenum line to the background is more favourable.

Analysing crystal   The diffraction grating used is usually a single crystal. A crystal is made up of layers of atoms or molecules arranged in a well-ordered system. An impinging beam of X-rays is reflected at each layer in the crystal. By varying the angle of incidence a single X-ray wavelength can be selected. The wavelength obtained is given by the Bragg's equation

$$n\lambda = 2d \sin \theta$$

where,

d is the spacing of atomic layers parallel to the crystal surface, and

θ is the angle of incidence of the X-ray beam with the crystal.

Such a crystal which splits the beam of X-rays into the component wavelengths is called an analysing crystal.

Detectors

Ionization counters   Consider a cylinder of suitable material placed in the centre of a positively charged wire. The cylinder can be sealed and filled with filler gas like helium. If an X-ray photon enters the cylinder, it will collide with and ionize a molecule of the filler gas, creating a primary ion pair. The electron is attracted to the centre of the wire by its positive charge. With no voltage applied, the electron and the positive ion recombine and no current flows. As the voltage is slowly increased, an increasing number of electrons reach the centre. a current is produced which is independent of small changes in voltage. A detector operating under these voltage conditions is called an ionizing counter.

Proportional counters   As the voltage increases further, the electrons moving towards the centre of the wire are increasingly accelerated until some of them have sufficient energy to ionize other atoms of filler gas. Thus secondary ion pairs are produced. More electrons reach the wire and an increased signal is obtained. This is the basis of proportional counters (Figure 10.3).

Gas flow proportional counters are used mainly for detection of longer wavelengths. Gas flows through it continuously. Where there are multiple detectors, the gas is passed through them in series, then led to waste. The gas is usually 90% argon and 10% methane although the argon may be replaced with neon or helium where very long wavelengths (over 5 nm) are to be detected. The argon is ionized by the incoming X-ray photons, and the electric field multiplies this charge into a measurable pulse. The methane suppresses the formation of fluorescent photons caused by recombination of the argon ions with stray electrons. The anode wire is typically tungsten or nichrome of 20–60 μm diameter. Since the pulse strength obtained is essentially proportional to the ratio of the detector chamber diameter to the wire diameter, a fine wire is needed. It must be strong enough to be maintained under tension so that it remains precisely straight and concentric with the detector. The window needs to be conductive, thin enough to transmit the X-rays effectively, but thick and strong enough to minimize diffusion of the detector gas into the high vacuum of the monochromator chamber. The materials often used are beryllium metal and aluminized polypropylene.

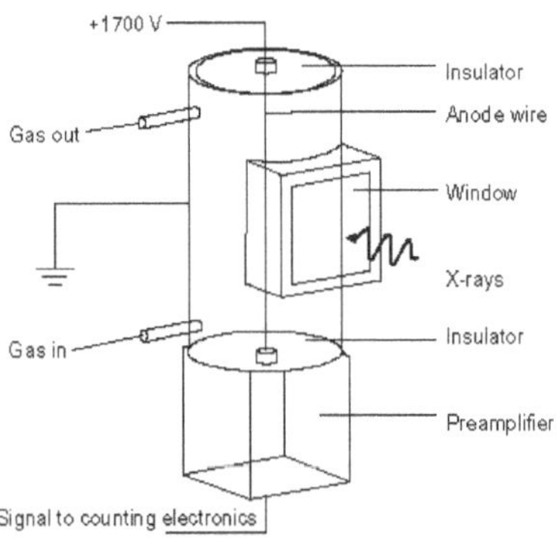

Figure 10.3   Proportional counter

Geiger counters   As the voltage is further increased, electrons formed in primary and secondary ion pairs are accelerated sufficiently to cause the formation of more ion pairs, producing plasma. This results in avalanche of electrons reaching the centre wire. The signal once generated is self-sustaining and is independent of intensity of X-ray falling on the detector. This is the basis of Geiger counter.

Scintillation counters   Scintillation counters consist of a scintillating crystal (typically of sodium iodide doped with thallium) attached to a photomultiplie. The crystal produces a group of scintillations for each photon absorbed, the number being proportional to the photon energy. This translates into a pulse from the photomultiplier of voltage proportional to the photon energy. The crystal must be protected with a relatively thick aluminium/beryllium foil window, which limits the use of the detector to wavelengths below 0.25 nm. Scintillation counters are often connected in series with a gas flow proportional counter: the latter is provided with an outlet window opposite the inlet, onto which the scintillation strikes.

Semiconductor detectors   Si(Li) and Ge(Li) are generally used as detectors. In this system, a very pure silicon block is set up with a thin film of lithium metal plated on to one end. The intrinsic material is light sensitive and if it is exposed to X-rays, an electron–hole pair is formed. Under the influence of an applied voltage, the electrons move towards the positive charge and the holes towards the negative charge. The voltage generated is a measure of the X-ray intensity falling on the crystal. This process is analogous to that of the proportional counter. When reached the lithium coating, a pulse is generated. The number of pulses is a direct measure of the number of X-ray photons falling onthe detector.

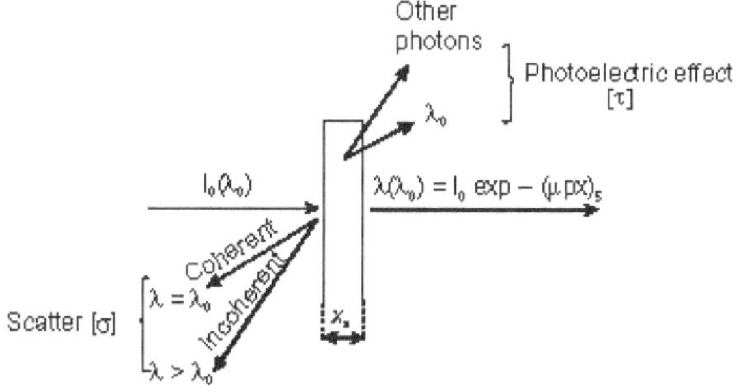

Figure 10.4   Interaction of X-ray photon with matter

when this happens, the wavelength of the transmitted beam is unchanged and the intensity of the transmitted beam I ($\ddot{e}_0$) is given by equation (1).

$$I(\lambda_0) = I_0 \exp-(\mu_a \rho_a x_a) \qquad (1)$$

where, μ_a is the mass attenuation coefficient of absorber a for the wavelength λ_0. It will be apparent from equation 1 that a number of photons equal to $(I_0 - I)$ have been lost in the absorption process, most of this loss being due to the photoelectric effect (Figure 10.4). Photoelectric absorption, usually designated t, will occur at each of the energy levels of the atom. Thus the total photoelectric absorption will be determined by the sum of each individual absorption within a specific shell. The value of mass attenuation μ_a referred to in Equation (1) is a function of both the photoelectric absorption and the scatter and given by equation (2).

$$\mu = f(t) + f(\sigma) \qquad (2)$$

However, f(t) is usually large in comparison with f(s). Because the photoelectric absorption is made up of absorption in the various atomic levels, it is an atomic number-dependent function. A plot of μ against λ contains a number of discontinuities called absorption edges, at wavelengths corresponding to the binding energies of the electrons in the various subshells. The absorption discontinuities are the major source of non-linearity between X-ray intensity and composition in both X-ray fluorescence (XRF) and X-ray diffraction (XRD).

X-ray diffraction (Scattering)

Scatter, s, will occur when an X-ray photon collides with one of the electrons of the absorbing element. Where this collision is elastic, i.e., when no energy is lost in the collision process, the scatter is said to be coherent (Rayleigh) scatter. X-ray diffraction is a special case of coherent scatter, where the scattered photons interfere with each other. It may happen that the scattered photon gives up a small part of its energy during the collision, especially where the electron with which the photon collides is only loosely bound. In this instance, the scatter is said to be incoherent (Compton scatter), and the wavelength of the incoherently scattered photon will be greater than λ_0.

X-ray fluorescence

When an element is placed in a beam of X-rays, absorption of X-rays takes place. The absorbing atoms become excited and emit X-rays of characteristic wavelength. This process is called X-ray fluorescence (XRF). Since the wavelength of the fluorescence is characteristic of the element being excited, measurement of this wavelength enables one to identify the fluorescing element. The intensity of the fluorescence depends on how much of that element is in the X-ray beam.

X-ray fluorescence (XRF) can be considered in a simple three-step process occurring at the atomic level:

1. An incoming X-ray knocks out an electron from one of the orbitals surrounding the nucleus within an atom of the material.

2. A hole is produced in the orbital, resulting in a high-energy unstable configuration for the atom.

3. To restore equilibrium, an electron from a higher energy outer orbital falls into the hole. Since this is a lower energy position, the excess energy is emitted in the form of a fluorescent X-ray.

The energy difference between the expelled and replacement electrons is characteristic of the element atom in which the fluorescence process is occurring. Thus, the energy of the emitted fluorescent X-ray is directly linked to a specific element being analysed. It is this key feature which makes XRF such a fast analytical tool for elemental composition.

In general, the energy of the emitted X-ray for a particular element is independent of the chemistry of the material. For example, a calcium peak obtained from $CaCO_3$, CaO and $CaCl_2$ will be in exactly the same spectral position for all three materials.

Since most atoms comprise a number of electron orbitals (e.g. K shell, L shell, M shell), a number of possible fluorescent transitions are possible.

For example, interaction of X-rays with an atom having K, L and M shells could result in a hole forming in the K shell, which is then filled by an electron from the L shell or from the M shell. In either case, these are termed K transitions. Alternatively, a hole could be formed in the L shell, subsequently filled by an electron from the M shell (termed an L transition).

Thus, for a single element, a number of XRF peaks are possible, and typically these will all be present in the spectrum, with varying intensities. They form a characteristic fingerprint for a specific element.

Specimen Preparation Techniques

Because X-ray spectrometry is essentially a comparative method of analysis, it is vital that all standards and unknowns be presented to the spectrometer in a reproducible and identical manner. Any method of specimen preparation must give specimens that are reproducible and which, for a certain calibration range, have similar physical properties including mass attenuation coefficient, density, particle size and particle homogeneity. In addition, the specimen preparation method must be rapid and cheap and must not introduce extra significant systematic errors, for example, the introduction of trace elements from contaminants in a diluent. Specimen preparation is an important factor in the ultimate accuracy of any X-ray determination, and many papers have been published describing a multitude of methods and recipes for sample handling. In general, samples fit into three main categories:

1. Samples that can be handled directly following some simple pretreatment such as pelletizing or surfacing, e.g. homogeneous samples of powders, bulk metals or liquids.

2. Samples that require significant pretreatment, e.g. heterogeneous samples, samples requiring matrix dilution to overcome inter-element effects, and samples exhibiting particle size effects.

3. Samples that require special handling treatment, e.g. samples of limited size, samples that require concentration or prior separation, and radioactive samples.

An ideal specimen for XRF analysis is the one in which the analysed volume of specimen is representative of the total specimen, which is, by itself, representative of the sample submitted for analysis. There are many forms of specimens suitable for XRF analysis, and the form of the sample as received will generally determine the method of pretreatment. It is convenient to refer to the material received for analysis as the sample, and that which is actually analysed in the spectrometer as the specimen. While the direct analysis of certain materials is certainly possible, more often than not, some pretreatment is required to convert the sample to the specimen. This step is referred to as specimen preparation. In general, the analysts would prefer to analyse the sample directly, because if it is taken as received any problems arising from sample contamination that might occur during pretreatment can be avoided. In practice, however, there are three major constraints that may prevent this ideal circumstance from being achieved:

i.   sample size,

ii. sample size homogeneity and

iii. sample composition heterogeneity.

Problems of sample size are frequently severe in the case of bulk materials such as metals, large pieces of rock, etc. Problems of sample composition heterogeneity will generally occur under these circumstances as well, and in the analysis of powdered materials, heterogeneity must almost always be considered. The sample as received may be either homogeneous or heterogeneous; in the latter case, it may be necessary to render the sample homogeneous before an analysis can be made. Heterogeneous bulk solids are generally the most difficult kind of sample to handle, and it may be necessary to dissolve or chemically treat the material in some way to give a homogeneous preparation. Heterogeneous powders are either ground to a fine particle size and then pelletized, or fused with a glass-forming material such as borax. The solid material in liquids or gases must be filtered out and the filter is analysed as a solid. When the analyte concentrations in liquids or solutions are too high or too low, dilution or preconcentration techniques may be employed to bring the analyte concentration within an acceptable range.

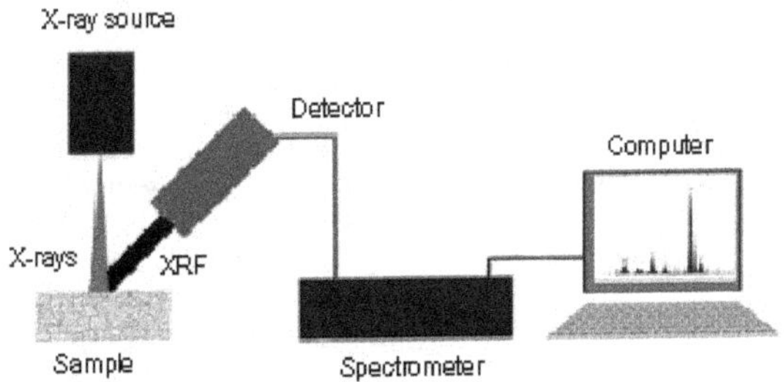

Figure 10.5   Components of XRF spectrometer

The key components of a typical XRF spectrometer (Figure 10.5) are

1.   Source of X-rays used to irradiate the sample.
2.   Sample.
3.   Detection of the emitted fluorescent X-rays.

The resulting XRF spectrum shows the intensity of X-rays (usually in counts per second) as a function of energy (usually in eV).

Wavelength-dispersive and energy-dispersive spectrometers

X-ray spectrometers fall roughly into two categories: wavelength-dispersive instruments and energy-dispersive instruments. The wavelength-dispersive system was commercially introduced in early 1950s, and since mid-1970s has developed into a widely accepted analytical tool. Analytical chemists use a wide range of instruments for the qualitative and quantitative analysis of multi-element samples, and when choosing a technique, they will generally consider some factors such as sensitivity, speed, accuracy, cost, range of applicability, and so on. Within the two major categories of X-ray spectrometers specified, there is of course a wide diversity of instruments available but most of these will fit the four basic types:

- Simultaneous wavelength-dispersive
- Sequential wavelength-dispersive
- Bremsstrahlung source energy-dispersive
- Secondary target energy-dispersive.

These different types of instruments differ in the type of source used for excitation, the number of elements which they can measure at a time, the speed at which they collect data and finally, their price range. The instruments listed above are, in principle at least, capable of measuring all the elements in the periodic classification from $Z = 9$ and upwards. They can be fitted with multisample handling facilities and can be automated by using minicomputers. They have a precision of the order of a few tenths of one percent and also have sensitivities down to even low parts per million level. As far as the analyst is concerned, they differ only in their speed, cost and number of elements measurable at the same time. Single-channel wavelength-dispersive spectrometers are typically employed for both routine and non-routine analysis of a wide range of products, including ferrous and non-ferrous alloys, oils, slags and sinters, ores and minerals, thin films, and so on. These systems are very flexible but are somewhat slow compared to multichannel spectrometers. The multichannel wavelength-dispersive instruments are used solely for routine, high throughput, analyses where there is a great need for fast accurate analysis, and where flexibility is of no importance. Energy-dispersive spectrometers have the great advantage of being able to display information on all elements at the same time. They lack somewhat in resolution over the wavelength-dispersive systems but also find great application in quality control, troubleshooting, and so on. They have been particularly effective in the fields of scrap alloy sorting, in forensic science and in the provision of elemental data to supplement X-ray powder diffraction data.

Wavelength-Dispersive Spectrometer

In order to maintain the required geometric conditions, a goniometer is used to ensure that the angles between the source and crystal, and the crystal and detector, are kept the same. The output from a wavelength-dispersive spectrometer may be either analog or digital. For qualitative work, an **analog output** is traditionally used, and in this case a rate meter is used to integrate the pulses over short time intervals, typically of the order of a second or so. The output from the rate meter is fed to an x/t recorder that scans at a speed which is conveniently coupled with that of the goniometer. The recorder thus displays an intensity/time diagram, and subsequently an intensity/2θ diagram. Tables are then used to interpret the wavelengths. For quantitative work, it is more convenient to employ **digital counting**, and a timer/scaler combination is provided to allow the pulses to be integrated over a period of several tens of seconds and then to be displayed as count or count rate. Most modern wavelength-dispersive spectrometers are controlled in some way by a minicomputer or microprocessor and by the use of specimen changers and are capable of very high specimen throughput. Once set up, the spectrometers will run virtually unattended for several hours (Figure 10.6).

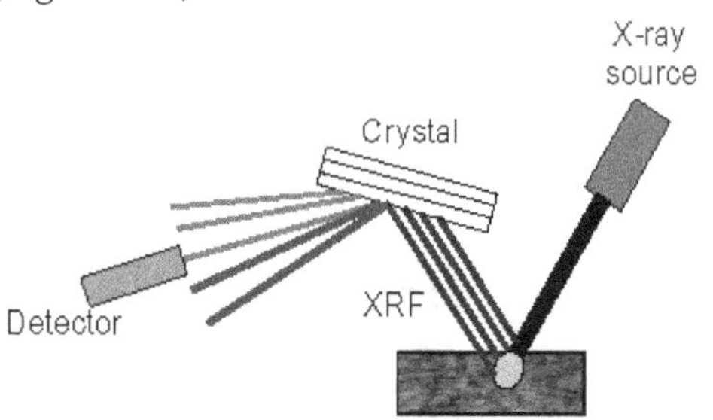

Figure 10.6   The energy-dispersive spectrometer

Energy-dispersive Spectrometer

Like the wavelength-dispersive spectrometer, the energy-dispersive spectrometer also consists of three basic units—excitation source, spectrometer and detection system. In this case, however, the detector itself acts as the dispersion agent. The detector generally employed is the Si(Li) detector. The Si(Li) detector consists of a small cylinder (about 1 cm diameter and 3 mm thick) of p-type silicon that has been compensated by lithium to increase its electrical resistivity.

A Schottky barrier contact on the front of the silicon disk produces a p-i-n type diode (one with p-doped, isolating and n-doped layers). In order to inhibit the mobility of lithium ions and to reduce electronic noise, the diode and its pre-amplifier are cooled to the temperature of liquid nitrogen. Incident X-ray photons interact to produce a specific number of electron–hole pairs. The charge produced is swept from the diode by the bias voltage to a charge-sensitive pre-amplifier. A charge loop integrates the charge on a capacitor to produce an output pulse. A pre-amplifier is responsible for collecting this charge on a feedback capacitor to produce a voltage pulse proportional to the original X-ray photon energy. Thus when a range of photon energies is incident upon the detector, an equivalent range of voltage pulses is produced as a detector output. A multichannel analyser is used to sort the arriving pulses at its input to produce a histogram representation of the X-ray energy spectrum. The output from an energy-dispersive spectrometer is generally displayed on some sort of visual display unit. The operator is able to dynamically display the contents of the various channels as an energy spectrum and provision is generally made to allow zooming in on the portions of the spectrum of special interest, to overlay spectra, to subtract background, and so on, in a rather interactive manner. Similar to wavelength-dispersive systems, nearly all energy-dispersive spectrometers will incorporate some form of minicomputer, which is available for spectral stripping, peak identification, quantitative analysis, and a host of other useful functions.

Analytical Applications of X-ray Methods

X-ray absorption   If the wavelength of an X-ray beam is short enough, it will excite an atom that is in its path. According to thumb rule, the X-rays emitted from a particular element will be absorbed by elements with a lower atomic number. The ability of each element to absorb increases with increasing atomic number. For example, the human arm consists of flesh, blood and bone. The flesh is made up of carbon, nitrogen, oxygen and hydrogen. These are low-atomic-weight elements and so they do not absorb much X-rays. In contrast, bone contains large quantities of calcium and phosphorus and the atomic numbers of these elements are much higher. Hence their absorptive power is also high. When an X-ray picture of an arm is taken, only the image of the bone is obtained.

Two elements present in a sample will absorb the X ray beam to different degrees. This property is used to detect impurities, segregations and contours of different phases. It is also used to detect voids, or to segregate impurities such as oxides, in welds and joints. The presence of blowholes in an absorption photograph indicates that the weld is mechanically weak.

X-ray absorption methods are non-destructive and it is independent of the chemical state of the elements. Low sensitivity of this method makes it more useful for analysis of major constituents than for trace metals.

X-ray fluorescence   This method is used for elemental analysis particularly for the analysis of metals and non-metals with atomic number greater than 12. The intensity of fluorescence is independent of chemical state of the elements and hence chemical preparation prior to X-ray fluorescence is not necessary. This method is also non-destructive.

The fluorescence spectra are very simple, and overlap of emission lines from other elements is unlikely.

Qualitative analysis can be carried out by measuring the angle of diffraction of the fluorescent X-rays. From this, the wavelength of fluorescence can be calculated. Each element fluoresces at its characteristic wavelengths and hence identification can be done easily.

Quantitative analysis can be carried out by measuring the intensity of fluorescence.

For example, in mining and metallurgy, XRF is used for the analysis of ores, concentrates and drilled cores. In the rubber industry, determination of sulphur can be carried out by XRF.

REVIEW QUESTIONS

1. Draw a schematic diagram of an X-rays source and describe its operation.

2. What is the origin of K_α X-rays lines?

3. What are the three major analytical fields of X-ray spectroscopy? State three analytical applications of each field.

4. Can X-rays from a tungsten target be used to excite copper atoms? Can X-rays from a copper target be used to excite tungsten atoms? Explain.

5. State and explain X-ray fluorescence.

6. Discuss briefly absorption of X-rays by matter.

7. How are X-rays detected and measured?

8. The energy difference between L and K levels is larger than that between the M and L level. Why?

9. The X-ray targets of high atomic number elements are generally preferred in X-rays studies. Why?

10. Describe the specimen preparation techniques for X-ray fluorescence.

11. Explain wavelength-dispersive and energy-dispersive X-ray spectrometrs.

12. Give a detailed account of analytical applications of X-rays.

Chapter II

SURFACE-SENSITIVE SPECTROMETRY

Introduction

Recent years have seen a significant development in the study of surface or interfacial composition of solid specimens. Surface-sensitive analytical techniques may be grouped into two categories—imaging and compositional analytical techniques.

The imaging techniques such as atomic force microscopy (AFM) and scanning tunnelling microscopy (STM) provide an image of the surface. Auger electron spectroscopy (AES), X-ray photoelectron spectroscopy (XPS) and secondary ion mass spectrometry (SIMS) are elemental analysis techniques. They are used to determine the concentration of specific elements at the surface. They also provide chemical-bonding information.

Surface spectroscopy involves probing a target sample with a flux of energetic particles and detecting characteristic particles emitted from the surface, after interaction. The probe beam may be photons, electrons or ions. The energy-exchange interactions that give rise to signals may be effectively stimulated by a variety of energy sources. The activation may be either by the electromagnetic radiation as in XPS (also called electron spectroscopy for chemical analysis (ESCA)) or by a beam of incident ions as in SIMS, or by an incident electron beam as in AES.

There are two significant drawbacks in surface analysis, resulting from the thinness of the sampled region. First, the quantity of sample is always small, i.e., only a few atoms thick. Therefore the techniques are micro-analytical. Second the sample preparation methods are crucial to obtain a meaningful analysis. For example, if a thin layer (only 10 nanometres thick) of oil covers a metal sample, surface analysis may show only organic compounds in the sample instead of giving information on metal surface.

An approximate definition can be made with respect to the thickness of the surface to be analysed. It is in the range of five times the atomic or molecular diameter of the substance. This means that for an atomic solid such as iron or silicon, the surface is about 1 nm thick. For a molecular solid such as a polymer, the surface will be about five times the diameter of the monomer or 10 nm thick. Beneath that level, the solid materials possess bulk properties.

IMAGING TECHNIQUES

Scanning Tunnelling Microscopy

The scanning tunnelling microscope was developed by Gerd Binnig and Heinrich Rohrer in the early 1980s, a development that earned them the Nobel Prize for Physics in 1986.

The principle of scanning tunnelling microscopy (STM) is based on the tunnelling current between a metallic tip, which is sharpened to a single-atom point, and a conducting material. A small bias voltage (mV to 3 V) is applied between an atomically sharp tip and the sample. If the distance between the tip and the sample is large, no current flows. However, when the tip is brought very close ($\leq 10\text{Å}$) without physical contact, a current (pA to nA) flows across the gap between the tip and the sample (Figure 11.1). This current is called tunnelling current which is the result of the overlapping wave functions between the tip atom and surface atom.

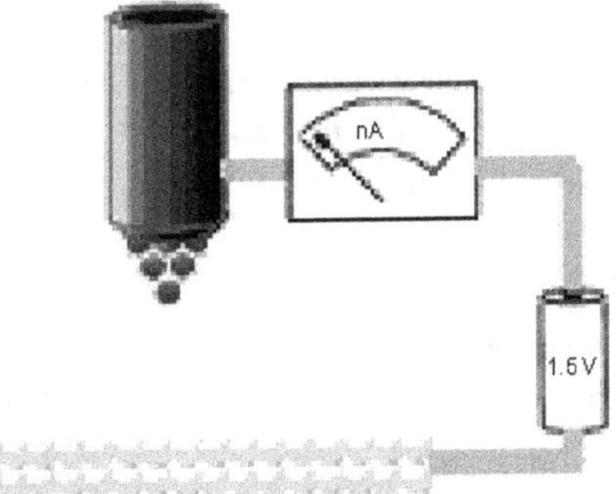

Figure 11.1   Scanning tunnelling microscope

Electrons can tunnel across the vacuum barrier separating the tip and the sample in the presence of a small bias voltage.

The magnitude of tunnelling current is extremely sensitive to the gap distance between the tip and the sample, to the local density of electronic states of the sample and to the local barrier height. The density of electronic states is the amount of electrons exit at specific energy. As we measure the current, with the tip moving across the surface, atomic information of the surface can be mapped out.

Tunnelling current is originated from the wavelike properties of particles (electrons, in the case) in quantum mechanics. When an electron is incident upon a vacuum barrier, with potential energy larger than the kinetic energy of the electron, there is still a non-zero probability that it may traverse the forbidden region and reappear on the other side of the barrier. It is shown by the leak out electron wave function in Figure 11.2.

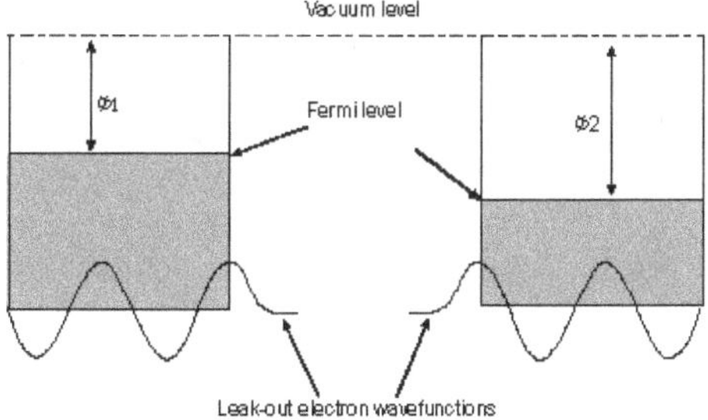

Figure 11.2   Diagram showing leak-out electron wave function

If two conductors are very close, their leak out electron wave functions overlap. The electron wavefunctions at the Fermi level have a characteristic exponential inverse decay length (K) given by

$$K = \frac{\sqrt{2m\Phi}}{\hbar}$$

(1)

where, m is the mass of electron, and Φ is the local tunnelling barrier height or the average work function of the tip and sample. When a small voltage, V, is applied between the tip and the sample, the overlapped electron wave functions permit quantum mechanical tunnelling, and a current (I) will flow across the vacuum gap (Figure 11.3).

At low voltage and temperature

$$I \propto \exp(-2Kd)$$

(2)

where, d is the distance between tip and sample. If the distance is increased by 1 Å, the current flow will decrease by an order of magnitude. So the sensitivity to vertical distance is terribly high. As the tip scans across the surface, it gives atomic resolution image.

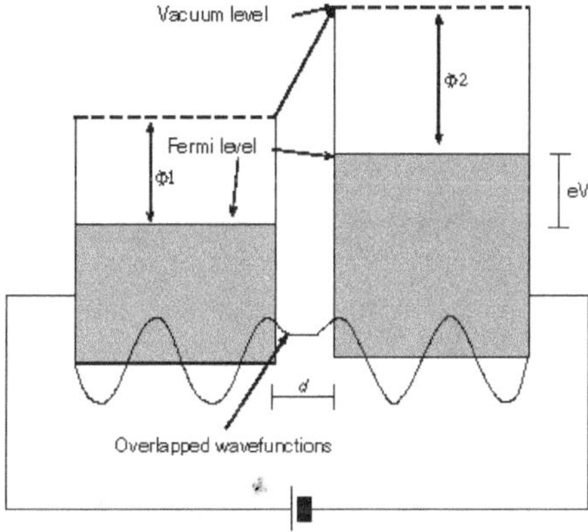

Figure 11.3   Tunnelling current

A piezo-electric scanner is used to accurately position an atomically sharp tip above a sample. Changing the position in the lateral (x, y) plane allows to scan continuously across the sample surface and changing the vertical (z) position allows to maintain desired tip–sample distance. If that distance becomes small enough (10–0.1 Å) and voltage is applied between the sample and the tip, the tunnelling current (1–0.1 nA) can be observed. This current depends exponentially on the tip–sample separation. Therefore, if a feedback loop is used to adjust the vertical position to keep the current constant (constant current scanning mode), tip–sample separation can be kept constant with great precision. Alternatively the z-coordinate can be held constant and the tunnelling current recorded. Since the current is essentially proportional to the density of electronic states in the sample, the first method maps the constant density of states contours and the second method maps the actual density of states (Figure 11.4).

The invention of STM has left a great impact on surface science. The use of STM to study the surface of metals and semiconductors can provide non-trivial, real space information, especially in studying semiconductors such as Si (100) surface, which is the technologically important Si substrate material for microelectronic device fabrication. The STM image of Si (100) surface gives direct confirmation of dimers formation during surface reconstruction, although it has been previously suggested by theoretical calculations and other experimental observations.

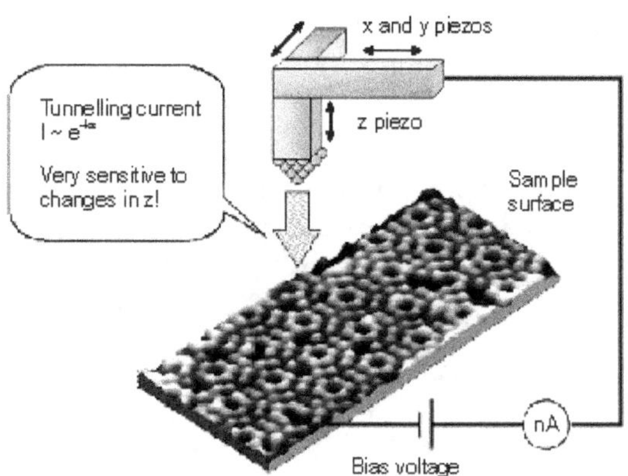

Figure 11.4   Schematic representation of basic STM set-up

Atomic Force Microscopy

The **atomic force microscope** (AFM) or scanning force microscope (SFM) is a very-high-resolution type of scanning probe microscope with demonstrated resolution of fractions of a nanometre. Binnig, Quate and Gerber invented the first AFM in 1986. The AFM is one of the foremost tools for imaging, measuring and manipulating matter at the nanoscale. The term **microscope** is actually a misnomer because it implies looking, while in fact the information is gathered by **feeling** the surface with a mechanical probe. Piezoelectric elements that facilitate tiny but accurate and precise movements on (electronic) command enable precise scanning.

In an AFM, a constant force is maintained between the probe and sample when the probe is scanned across the surface. The constant force is maintained by measuring the force with the **light lever sensor** and using a **feedback control** electronic circuit to control the position of the Z piezoelectric ceramic (Figure 11.5). By monitoring the motion of the probe during scanning, a three-dimensional image of the surface is constructed.

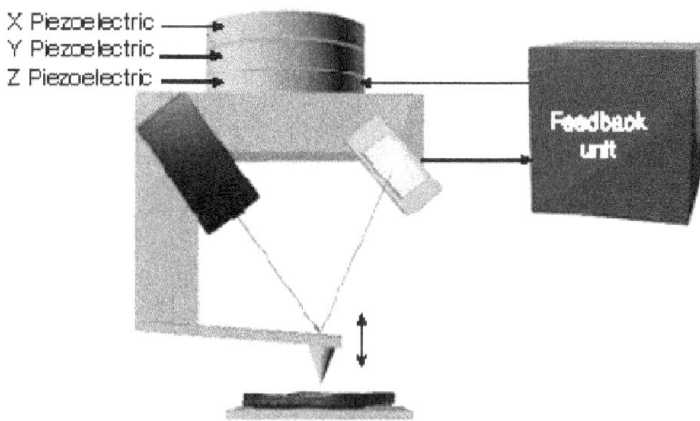

Figure 11.5 Primary components of the light lever atomic force microscope

The motion of the probe over the surface is generated by piezoelectric ceramics that move the probe and light lever sensor across the surface in X and Y directions. The X- and Y-piezo ceramics are used to scan the probe over the surface.

The principles on how the AFM works are very simple. An atomically sharp tip is scanned over a surface with feedback mechanisms that enable the piezoelectric scanners to maintain the tip at a constant force (to obtain height information), or at a constant height (to obtain force information) above the sample surface. Tips are typically made from Si_3N_4 or Si, and extended down from the end of a cantilever. The nanoscope AFM head employs an optical detection system in which the tip is attached to the underside of a reflective cantilever. A diode laser is focused on to the back of a reflective cantilever. As the tip scans the surface of the sample, moving up and down with the contour of the surface, the laser beam is deflected off the attached cantilever into a dual element photodiode. A photodetector measures the difference in light intensities between the upper and lower photodetectors, and then converts to voltage. The feedback from the photodiode difference signal, through software control from the computer, enables the tip to maintain either a constant force or constant height above the sample. In the constant force mode, the piezoelectric transducer monitors the real time–height deviation. In the constant height mode, the deflection force on the sample is recorded. The latter mode of operation requires calibration parameters for the scanning tip to be inserted in the sensitivity of the AFM head during force calibration of the microscope.

According to the interaction of the tip and the sample surface, the AFM can be classified as repulsive or contact mode and attractive or non-contact mode. The contact mode essentially operates as a low-load, high-resolution surface profiler. In the case of noncontact mode, the attractive force near contact plays a significant role. It works under ultra-high vacuum conditions.

Resolution in an Atomic Force Microscope

Traditional microscopes have only one measure of resolution: the resolution in the plane of an image. An atomic force microscope has two measures of resolution—the resolution in the plane of the measurement (in-plane) and the resolution in the direction perpendicular to the surface (vertical).

In-plane resolution The in-plane resolution depends on the geometry of the probe that is used for scanning. In general, the sharper the probe, the higher is the resolution of the AFM image. The theoretical line scans of two spheres that are measured with a sharp probe and a dull probe are shown in Figure 11.6.

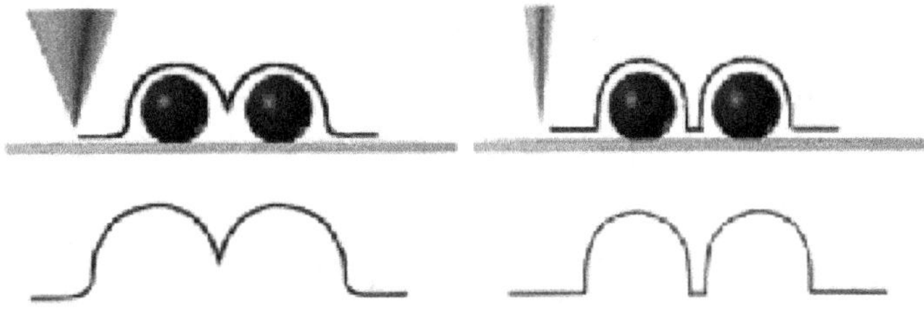

Figure 11.6 Theoretical line scans of two spheres using dull and sharp probes

The image on the right will have a higher resolution because the probe used for the measurement is much sharper.

Vertical resolution The vertical resolution in an AFM is established by relative vibrations of the probe over the surface. The sources for these vibrations are acoustic noise, floor vibrations, and thermal vibrations. A maximum vertical resolution is obtained by minimizing the vibrations of the instrument.

The atomic force microscope (AFM) is used to solve processing and materials problems in a wide range of technologies affecting the electronics, telecommunications, biological, chemical, automotive, aerospace, and energy industries. The materials that can be investigated include thin and thick film coatings, ceramics, composites, glasses, synthetic and biological membranes, metals, polymers, and semiconductors. It is also used to study phenomena such as abrasion, adhesion, cleaning, corrosion, etching, friction, lubrication, plating, and polishing. By using AFM one cannot only image the surface in atomic resolution but also measure the force at nano-newton scale.

COMPOSITIONAL ANALYTICAL TECHNIQUES

Auger Electron spectroscopy

Auger Electron Spectroscopy (**Auger spectroscopy** or AES) was developed in the late 1960s. It is named after Pierre Auger, a French Physicist who first observed the effect in the mid-1920s. It is a surface-specific technique utilizing the emission of low-energy electrons in the **Auger process** and is one of the most commonly employed surface analytical techniques for determining the composition of the surface layers of a sample.

Auger spectroscopy involves three basic steps:

1. Atomic ionization (by removal of a core electron)
2. Electron emission (the Auger process)
3. Analysis of the emitted Auger electrons

In Auger spectroscopy, the excitation source is a finely focused electron beam. Upon sample bombardment, a transfer of energy occurs which excites a core electron into an orbital of higher energy. Once in this excited state, the atom has two possible modes of relaxation: emission of an X-ray, or emission of an Auger electron. In both the processes, the emitted particle will have an energy characteristic of the parent element. The last stage is simply a technical problem of detecting charged particles with high sensitivity, with an additional requirement that the kinetic energies of the emitted electrons must be determined.

An energy spectrum of the detected electrons shows peaks assignable to the elements present. The ratio of the intensities of Auger electron peaks can provide a quantitative determination of surface composition. The Auger multiprobe is capable of producing elemental composition spectra, surface images, selective elemental line scans and maps, and depth profiles. Imaging is achieved through detection of secondary, backscattered and Auger electrons. Elemental line scans detect concentrations along a line, while a map, which is considerably more time-consuming, detects elemental distribution over an area. Auger analysis involves only the topmost 20 Å, making it an extremely surface-sensitive technique.

Depth profiles are obtained by employing a controlled sputtering process which enables elemental concentration to be plotted as a function of depth. **Sputtering** is a process in which an ion gun is used to remove a few angstroms of the topmost surface of a sample. Sputtering and analysis are alternated until the desired depth is reached. The thickness in layered samples is calibrated by comparing the sputtering rate of a known thickness of SiO_2. Auger depth profiling is well suited for conductive samples composed of thin multilayers where thickness or interface properties must be determined.

Electrons Emitted from Samples

Figure 11.7 illustrates the energy levels of electrons in an isolated, multi-electron atom, with the conventional chemical nomenclature presented on the right side for these orbitals given on the right hand side.

However, scientists working with X-rays tend to use the alternative nomenclature on the left side (Figure 11.7) and it is this nomenclature that is used in Auger spectroscopy. The designation of levels to the shells (K, L, M, …)is based on their principal quantum numbers and therefore designated as 1, , 3, … respectively.

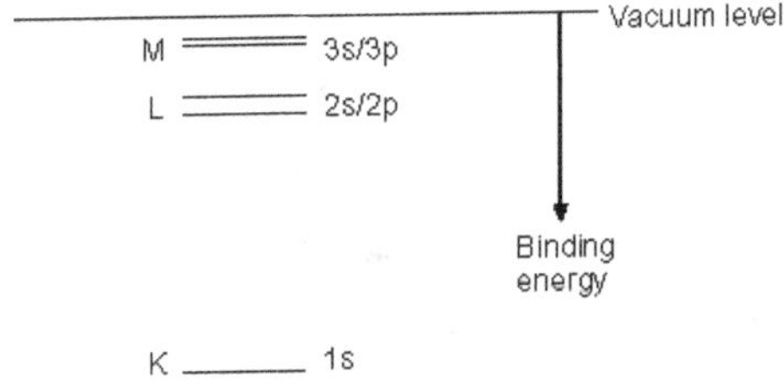

Figure 11.7   Energy levels of electrons

It is convenient to expand the part of the energy scale close to the vacuum level in order to more clearly distinguish between the higherlevels (Figure 11.8).

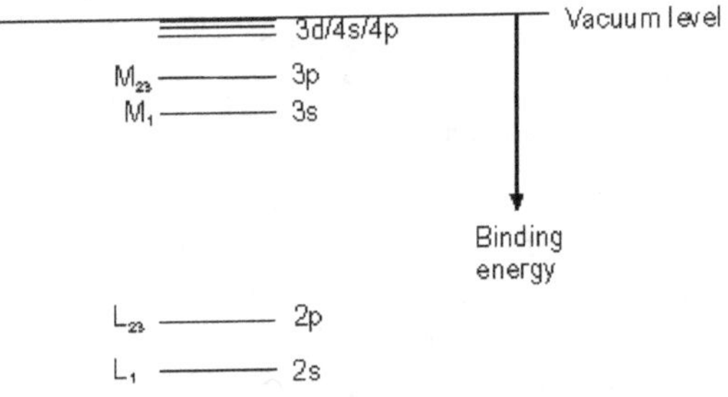

Figure 11.8   Auger nomenclature of energy levels of electrons

The numerical component of the Auger nomenclature (KLM style) is usually written as a subscript immediately following the main shell designation.

Levels with a non-zero value of the orbital angular momentum quantum number (1 > 0), i.e., p, d, f,. levels, show spin-orbit splitting. The magnitude of this splitting, however, is too small to be evident on this Figure 11.8. Hence, the double subscript for these levels, i.e., $L_{2,3}$ represents both the L_2 and L_3 levels).

In the solid state, the core levels of atoms are little perturbed and essentially remain as discrete, localized (i.e., atomic-like) levels. The valence orbitals, however, overlap significantly with those of neighbouring atoms, generating bands of spatially delocalized energy levels. Therefore, the energy- level diagram for the solid closely resembles that of the corresponding isolated atom, except for the levels closest to the vacuum level. Figure 11.9 shows the electronic structure of Na metal.

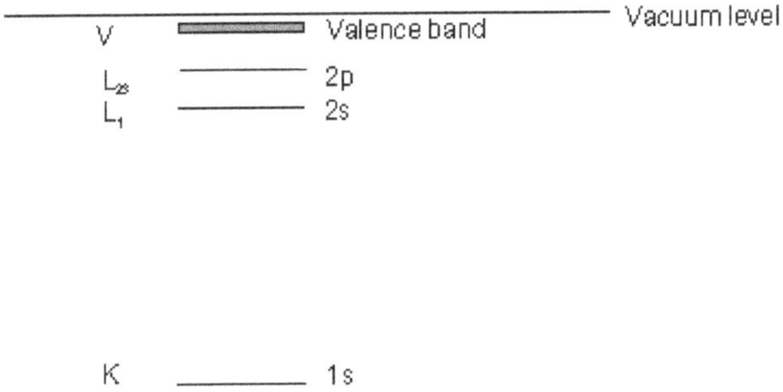

Figure 11.9   Auger nomenclature for sodium metal

The Auger Process and Auger Spectroscopy

In the following section, the Auger process is illustrated using the K, L_1 and $L_{2,3}$ levels. These could be the inner core levels of an atom in either a molecular or solid-state environment.

Ionization   The Auger process is initiated by creating a core hole. This is typically carried out by exposing the sample to a beam of high-energy electrons (typically having a primary energy in the range 2–10 keV). Such electrons have sufficient energy to ionize all levels of the lighter elements, and higher core levels o the heavier elements.

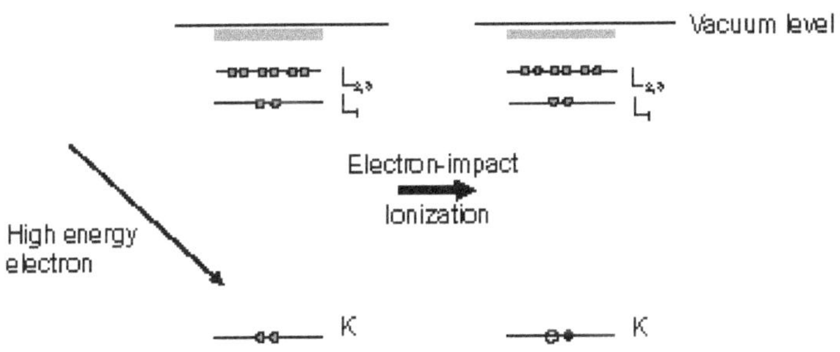

Figure 11.10   Ionization at K level

In Figure 11.10, ionization is shown to occur by removal of a K-shell electron, but in practice such a crude method of ionization will lead to ions with holes in a variety of inner shell levels. In some studies, the initial ionization process is instead carried out using soft X-rays ($h\nu$ = 1000–2000 eV). In this case, the acronym XAES is sometimes used. This change in the method of ionization has no significant effect on the final Auger spectrum.

Relaxation and auger emission   The ionized atom that remains after the removal of the core hole electron is, of course, in a highly excited state and will rapidly relax back to a lower energy state by one of the two routes:

1.   X-ray fluorescence

2.   Auger emission

Let us consider only the latter mechanism, an example of which is illustrated in Figure 11.11.

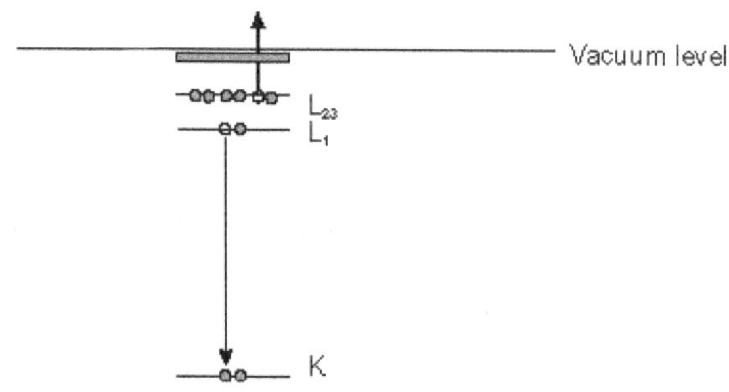

Figure 11.11   Auger emission

In Figure 11.11, an electron falls from a higher level to fill an initial core hole in the K-shell and the energy liberated in this process is simultaneously transferred to a second electron. A fraction of this energy is required to overcome the binding energy of this second electron, and the remaining energy is retained as kinetic energy (KE) by this emitted **Auger electron**. In the Auger process illustrated, the final state is a doubly ionized atom with core holes in the L_1 and $L_{2,3}$ shells.

We can make a rough estimate of the KE of the Auger electron from the binding energies of the various levels involved. In this particular example,

$$KE = (E_K - E_{L1}) - E_{L2,\,3}$$

Note   The KE of the Auger electron is independent of the mechanism of initial core hole formation.

The expression for the energy can also be rewritten in the form:

$$KE = E_K - (E_{L1} + E_{L2,\,3})$$

Therefore, it is clear from this expression that the latter two energy terms could be interchanged without any effect, i.e., it is actually impossible to say which electron fills the initial core hole and which is ejected as an Auger electron. They are indistinguishable.

An Auger transition is therefore characterized primarily by

i.   the location of the initial hole

ii.   the location of the final two holes

Although the existence of different electronic states (terms) of the final doubly-ionized atom may lead to fine structure in high-resolution spectra.

When describing the transition, the location of the initial hole is given first, followed by the locations of the final two holes in the order of decreasing binding energy, i.e., the transition illustrated is a $KL_1L_{2,3}$ transition.

If we consider these three electronic levels, there are clearly several possible Auger transitions: specifically,

$$K\,L_1\,L_1 \qquad\qquad K\,L_1\,L_{2,3} \qquad\qquad K\,L_{2,3}\,L_{2,3}$$

In general, since the initial ionization is non-selective and the initial hole may therefore be in various shells, there will be many possible Auger transitions for a given element—some weak and some strong in intensity. Auger spectroscopy is based upon the measurement of the kinetic energies of the emitted electrons. Each element in a sample being studied will give rise to a characteristic spectrum of peaks at various kinetic energies.

Figure 11.12 is an Auger spectrum of Pd metal, generated using a 2.5-keV electron beam to produce the initial core vacancies and hence to stimulate the Auger emission process. The main peaks for palladium occur between 220 and 340 eV. The peaks are situated on a high background that arises from the vast number of the so-called **secondary electrons** generated by a multitude of inelastic scattering processes.

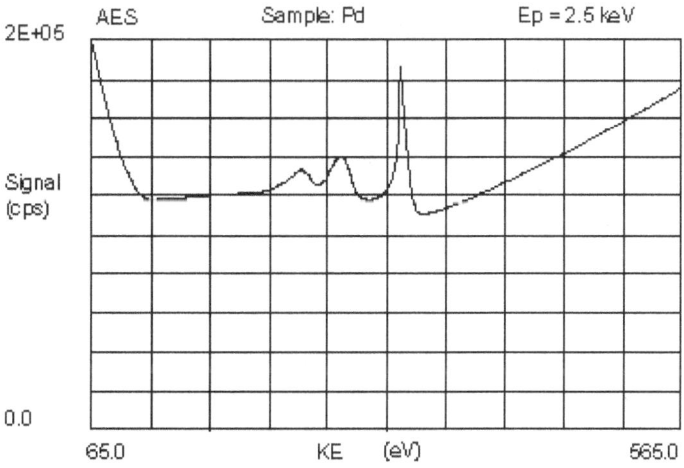

Figure 11.12   Auger spectrum of Pd metal

Auger spectra are also often shown in a differentiated form. It is possible to actually measure spectra directly in this form and by doing so can get a better sensitivity for detection. Figure 11.13 shows the same spectrum in such a differentiated form.

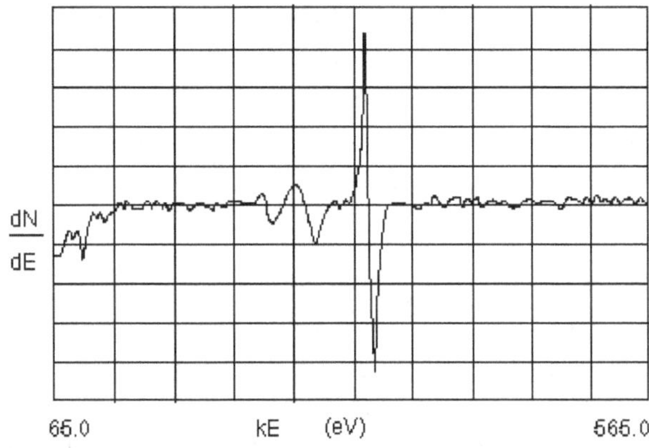

Figure 11.13   Auger spectrum of Pd metal in differentiated form

Uses of AES

Auger electron spectroscopy is used for elemental analysis of surfaces. It offers a high sensitivity (typically ca. 1% monolayer) for all elements except H and He. It serves as a means of monitoring surface cleanliness of samples. It provides compositional information for many types of surfaces, thin films, and interfaces. Typical samples include both raw semiconductor materials and finished electronic devices. Many of these devices consist of thin layers. For example, AES can distinguish between Si, SiO_2, SiO, and Si_3N_4 in a 10 nm layer on a silicon wafer. AES analytical volumes down to about 3×10^{-19} cc are possible. It is common to analyse individual small features within finished or partly finished electronic devices. Many other analyses rely on this microanalytical capability for characterizing heterogeneous materials. The fractured surfaces of a broken piece of steel might be examined for the presence of unusual elements such as lead at metal grain boundaries.

In addition, this basic technique has also been adapted for use in

1. Auger Depth Profiling to provide quantitative compositional information as a function of depth below the surface.

2. Scanning Auger Microscopy (SAM) to provide spatially resolved compositional information on heterogeneous samples.

Limitations of AES

Although widely useful, AES has some limitations (1) charging up of insulative surfaces when struck by the primary electron beam (2) damage to certain materials, especially organic ones, when struck by the electron beam (3) occurrence of matrix effects, i.e., signal alterations when some elements are present in particular matrices (4) cannot detect hydrogen or helium. (5) cannot provide for non-destructive depth profiles (6) requirement of samples that are small and compatible with high vacuum.

X-Ray Photoelectron Spectroscopy

Electrons can be emitted from atoms and molecules upon irradiation with light of sufficiently high energy. This is called photoemission and the electrons emitted are called photoelectrons. The energies of the electrons are related to the composition and structure of the sample. Photoelectron spectroscopy utilizes photo-ionization and energy-dispersive analysis of the emitted photoelectrons to study the composition and electronic state of the surface region of a sample.

When UV radiation is used, the technique is referred to as UPS (ultraviolet photoelectron spectroscopy) and the electrons emitted are referred to as the valence electrons of atoms on or close to the surface. If X-ray radiation is used, the technique is called XPS and this is sufficiently energetic to cause the emission of core electrons; further, since the X-ray can penetrate into the sample, the emitted electrons may arise from deeper in the bulk material compared with UPS.

The energy of an incident photon, $h\nu$, is related to the kinetic energy of ejected electrons as follows:

$$h\nu = E_{KE} + BE$$

where, BE is the binding energy of the level from which the electron was emitted.

For both XPS and UPS, the kinetic energy of the ejected electrons is measured using a hemispherical analyser. The basic requirements for a photoemission experiment (XPS or UPS) are

1. a source of fixed-energy radiation (an X-ray source for XPS or, typically, a He discharge lamp for UPS)

2. an electron energy analyser (which can disperse the emitted electrons according to their kinetic energy, and thereby can measure the flux of emitted electrons of a particular energy)

3. a high vacuum environment (to enable the emitted photoelectrons to be analysed without interference from gas-phase collisions)

Such a system is illustrated schematically in Figure 11.14.

There are many different designs of electron energy analyser but the preferred option for photoemission experiments is a concentric hemispherical analyser (CHA). It uses an electric field between two hemispherical surfaces to disperse the electronsaccording to their kinetic energy.

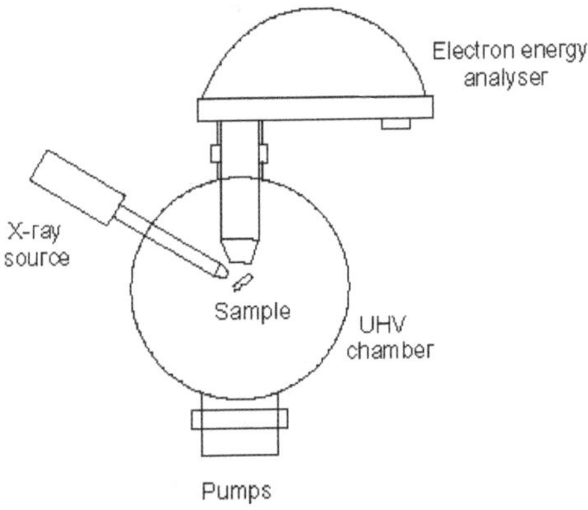

Figure 11.14 Block diagram of XPS

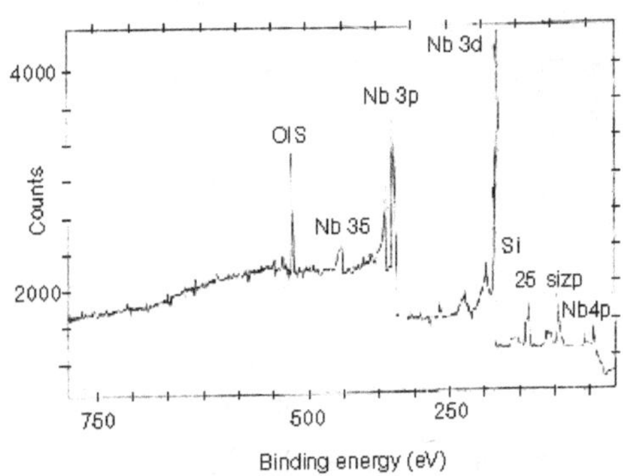

Figure 11.15 The XPS spectrum of niobium deposited on a silicon surface

The XPS spectrum of niobium deposited on silicon surface is shown in Figure 11.15. The major peaks below 500 eV can all be quite precisely correlated with the ionization energies of electrons from the various orbitals of niobium or silicon as the peak just above 500 eV arises from the 1s orbital of oxygen contaminating the surface. The sloping background is due to secondary electron emissions within the bulk of the sample. Clearly the technique is invaluable for qualitative analysis—the electrons being emitted come from deep within the atom concerned, well-shielded from any effect due to chemical bonding, and any atomic species near or on the surface can be instantly recognized from its ionization potential. Quantitative analysis is also possible, by observing the relative intensities from its ionization emission.

For each and every element, there will be a characteristic binding energy associated with each core atomic orbital, i.e., each element will give rise to a characteristic set of peaks in the photoelectron spectrum at kinetic energies determined by the photon energy and the respective binding energies.

The presence of peaks at particular energies therefore indicates the presence of a specific element in the sample under study. Furthermore, the intensity of the peaks is related to the concentration of the element within the sampled region. Thus, the technique provides a quantitative analysis of the surface composition. The XPS spectrum of Pd is drawn in Figure 11.16.

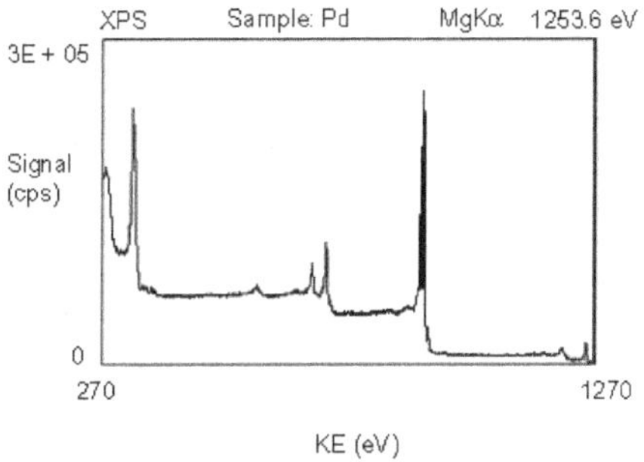

Figure 11.16   XPS spectrum of Pd

The main peaks occur at kinetic energies of 330, 690, 720, 910 and 920 eV.

Since the energy of the radiation is known, it is a trivial matter to transform the spectrum so that it is plotted against E as opposed to KE (Figure 11.17).

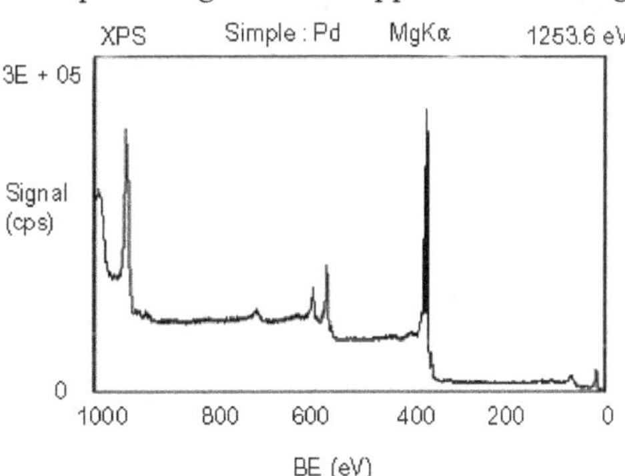

Figure 11.17   Binding energy of Pd derived from XPS

The most intense peak is observed to occur at a binding energy of 335 eV.

Working downwards from the highest energy levels are arrive at the following conclusions.

1.   The valence band (4d, 5s) emission occurs at a binding energy of ca. 0–8 eV (if measured with respect to the Fermi level), or alternatively at ca. 4–12 eV (if measured with respect to the vacuum level).

2.   The emission from the 4p and 4s levels gives rise to very weak peaks at 54 and 88 eV respectively.

3.   The most intense peak at ca. 335 eV is due to the emission from the 3d levels of the Pd atoms, whilst the 3p and 3s levels give rise to the peaks at ca. 534/561 eV and 673 eV respectively.

4.   The remaining peak is not an XPS peak. It is an Auger peak arising from X-ray induced Auger emission. It occurs at a kinetic energy of 330 eV (in this case, it is really meaningless to refer to an associated binding energy).

These assignmets are summarized in Figure 11.18.

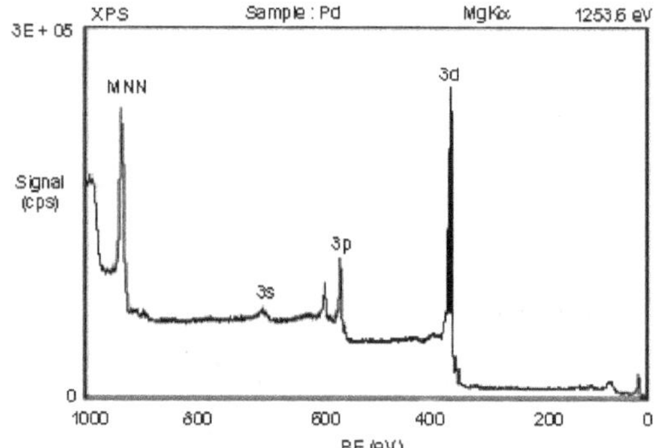

Figure 11.18   XPS of Pd with peak assignments

It may be further noted that

- there are significant differences in the natural widths of the various photoemission peaks, and

- the peak intensities are not simply related to the electron occupancy of the orbitals.

Secondary Ion Mass Spectrometry

Bombardment of a sample surface with a primary ion beam followed by mass spectrometry of the emitted secondary ions constitutes secondary ion mass spectrometry (SIMS). It is thus the mass analysis of positive and negative ions sputtered from the surface of a solid during ion bombardment. A spectrum of the intensity of the secondary ions versus mass is generated. The ion beam used to sputter the sample is called the primary ion beam and typically has a kinetic energy of 0.5 to 25 kilovolts. The sputter erosion process removes atomic and molecular ions and the ions produced at the surface for analysis make up the secondary ion beam.

Instrumentation

The basic components of SIMS include the primary ion source, secondary ion extraction optics, mass spectrometer and secondary ion detector.

Typical SIMS instruments use either a duoplasmatron or a surface ionization source (primary ion) or both.

The duoplasmatron (Figure 11.19) can function along with virtually any gas, but oxygen is the most common, because oxygen implantation into the sample surface enhances the ionization efficiency for electropositive elements. Before this oxygen enhancement effect was discovered, argon was commonly used. The oxygen plasma within the duoplasmatron source contains both O and O_2^+, and either can be extracted.

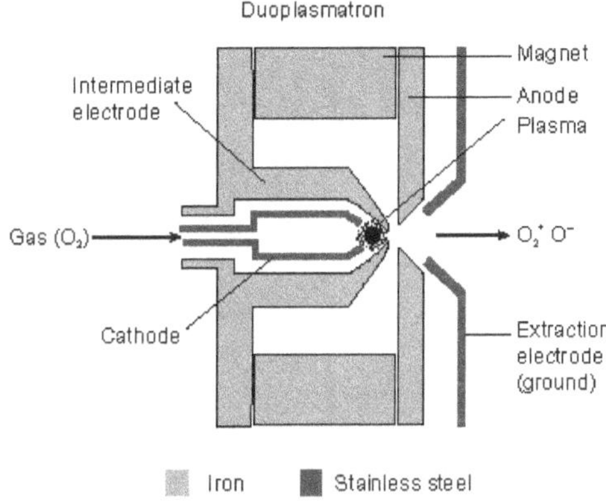

Figure 11.19   Parts of duoplasmatron

The caesium surface ionization source (Figure 11.20) produces Cs⁺ ions as Cs atoms vaporize through a porous tungsten plug.

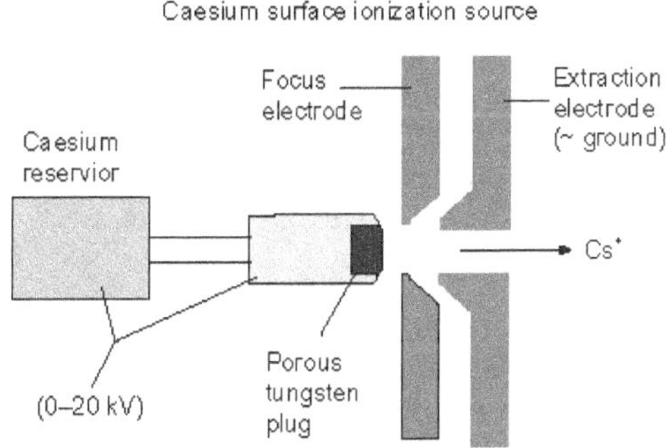

Figure 11.20   Caesium surface ionization source

The primary ions are extracted from the sources and are passed to the sample through the primary ion column. The column usually contains a primary beam mass filter that transmits only the ions with a specific mass-to-charge (m/z) ratio.

This mass filter eliminates the impurity species in the beam. For example, Cr, Fe, and Ni ions sputter from stainless steel surfaces within a duoplasmatron. Without a primary beam mass filter, these metal contaminants deposit onto the sample surface, raising the detection limits for stainless steel elements.

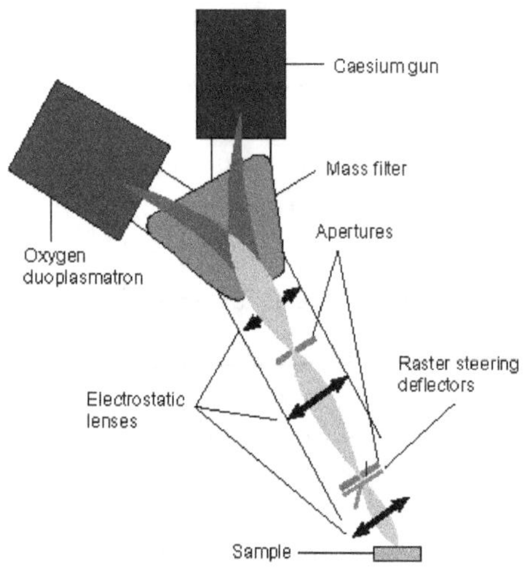

Figure 11.21   Mass filter

The electromagnetic active components are shown in thick black arrows (Figure 11.21). The ion beam trajectories (indicated in light grey) are greatly exaggerated in the lateral directions.

The electrostatic lenses and apertures control the intensity and width of the primary ion beam. Several aperture diameters are usually available at each aperture location. The intensity of the primary beam can be reduced by defocusing the ion beam onto the back of the first aperture (nearest the magnet). A narrow beam at the sample results from defocusing the ion beam with the middle lens onto the back of the second aperture, and then adjusting the last lens to transfer the image of the cross-over from behind the aperture onto the sample.

Electrostatic deflectors steer the primary beam in a raster pattern onto the sample. A finely focused primary ion beam delivers uniform primary beam intensity to an area on the sample. This leads to flat-bottomed sputter craters. The best depth resolution in a depth profile results when the secondary ions are sampled from the flat bottom of such a crater without contributions from the crater edges. Other deflectors (not shown) are located near the apertures. They help tune the primary beam through the middle of the electrostatic lenses.

Secondary ions are extracted from the sample as they are produced (Figure 11.22). If the components of a large mass spectrometer are held at ground potential, the sample must be held at high voltage, i.e., the accelerating potential. The secondary ions accelerate towards the ground plate of an electrostatic lens. This first lens is called the immersion or ion extraction lens. The second (transfer lens) focuses the ion beam onto entrance slits or aperture of the mass spectrometer. This two-lens system constitutes an ion microscope. The secondary ions could be projected on to an image detector for viewing the sample surface. Different transfer lenses produce different magnifications.

The electromagnetic active components are shown in light grey arrows (Figure 11.22) and the ion beam trajectories (indicated in dark grey) are greatly exaggerated in the lateral directions.

The field aperture is located approximately at the point where the ion beam image comes into focus. The entrance aperture is also called a contrast diaphragm. The smaller aperture diameters intercept ions with off-axis energy components. This not only reduces image aberrations but unfortunately it also reduces secondary ion intensity.

Ions that arise from off the secondary ion optical axis contribute to lower mass resolution. These off-axis ions arise because the primary beam raster pattern sputters an area rather than the single point where the axis intercepts the sample. The dynamic remittance deflectors adjust the secondary ion beam back on-axis. The deflectors operate in synchrony with the primary beam raster generator to provide continuous adjustment.

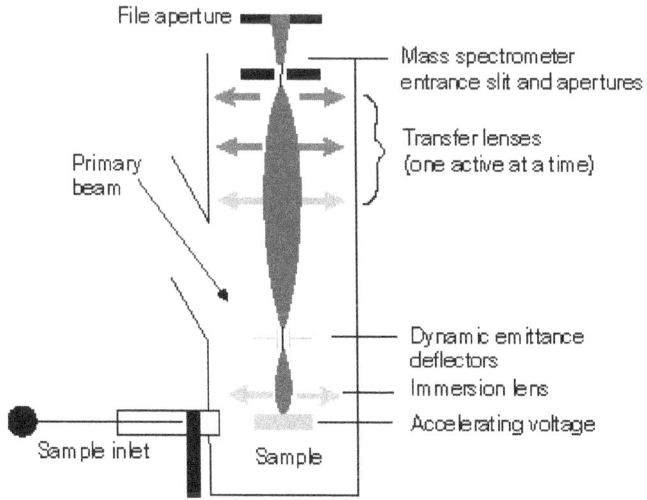

Figure 11.22   Extraction of secondary ions

Electrostatic energy analysers bend the lower energy ions more strongly than the higher energy ions. The sputtering process produces a range of ion energies. An energy slit can be set to intercept the high-energy ions (shown in light grey in Figure 11.23).

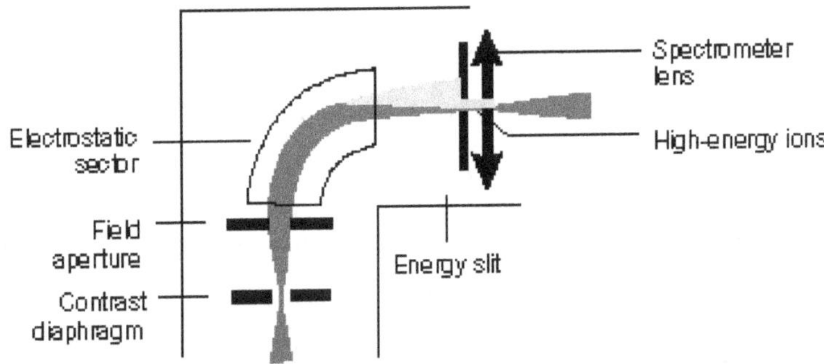

Figure 11.23 Electrostatic energy analyser

The electromagnetic active components are shown in black arrows. The ion beam trajectories (indicated in dark grey) are greatly exaggerated in the lateral directions.

Voltage offset is a strategy for enhancing monatomic ions over multiatomic. The monatomic ions have higher energy distributions. If the accelerating voltage is lowered (offset), more of the atomic ions still have enough energy to pass through the energy slits. In a typical SIMS experiment, the accelerating voltage is 4.5 kV, and the offset is 50 V. The inner jaw of the slits intercepts most (low energy) multiatomic ions. Both monatomic and multiatomic ion intensities are reduced in a voltage offset measurement, but multiatomic ions relatively more than monoatomic.

The inner and outer sector electrodes have voltages of opposite polarity. Their magnitude is about 10% of the ion-accelerating voltage. The ion image comes into focus, producing a virtual image inside the electrostatic sector behind the field aperture. The active surfaces of the electrostatic sector are spherical. This geometry transfers the image to the mass analyser with minimal distortion.

SIMS measures trace levels of all elements in the periodic table. SIMS also provides lateral and depth distributions (microanalysis) of these elements within a sample. The industries of electronic materials (semiconductors, optoelectric devices, etc.) are the largest users of SIMS. The geological community also uses SIMS for laterally resolved isotopic and elemental measurements.

REVIEW QUESTIONS

1. Why is flood gun used in ESCA–Auger equipment?

2. Why is ESCA sensitive to the chemical form of an element but Auger is not used for robtaining this information?

3. What is the relationship between Auger sensitivities and atomic number?

4. What sources are commonly used for generating ESCA and Auger spectra?

5. Explain scanning tunnelling microscopy.

6. Describe the technique of atomic force microscope.

7. Describe the different resolutions in atomic force microscope.

8. What in an Auger process?

9. Write a note on relaxation and Auger emission.

10. What is X-ray photoelectron spectroscopy?

11. Draw a schematic diagram of ESCA instrument?

12. Describe the functioning of electostatic energy analyser.

13. What is secondary ion mass spectrometry?

MULTIPLE CHOICE QUESTIONS

1. How many absorptions will the following compound have in its carbon NMR spectrum?

$$\begin{array}{c} H \\ \backslash \\ C = C \\ / \quad\quad \backslash \\ CH_3CH_2 \quad CH_3 \end{array} \quad \begin{array}{c} CH_3 \\ / \\ \end{array}$$

       (a) 3   (b) 4   (c) 5   (d) 6

2. Which of the following bonds would show the strongest absorption in the IR?

       (a) carbon–hydrogen   (b) oxygen–hydrogen

       (c) nitrogen–hydrogen   (d) sulphur–hydrogen

3. Proton NMR is useful for investigating the structure of organic compounds because

       (a)   organic compounds contain carbon atoms

       (b)   organic compounds are mostly covalent

       (c)   hydrogen atoms are found in nearly all organic compounds

       (d)   organic compounds have low boiling points

4. Which one of the following methods would be best for finding the identity of an organic compound?

       (a)   finding the m/z value of the molecular ion in its mass spectrum

       (b)   its proton NMR spectrum

       (c)   comparing its infrared spectrum with known examples

       (d)   measuring its melting point

5. The molecule $HOCH_2CH_2OH$ will have an NMR spectrum consisting of

       (a) two singlets   (b) a triplet and a doublet

       (c) two doublets   (d) a singlet and a doublet

6. In a triplet, the relative peak areas are in the ratio

       (a) 1: 1: 1   (b) 1 : 2 : 1   (c) 1: 3: 1   (d) 1: 4: 1

7. What is the multiplicity expected in the hydrogen NMR spectrum for the hydrogen atoms marked by a "star" in the following compound?

$$CH_3-\overset{\overset{\displaystyle O}{\|}}{C}-\overset{*}{C}H_2-CH_3$$

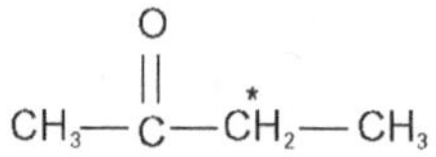

   (a) singlet   (b) triplet   (c) quartet   (d) heptet

8. What is the λ_{max} for the following compound?

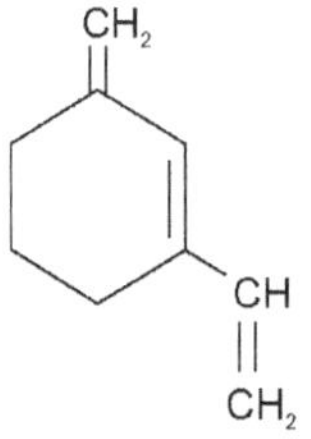

   (a) 229 nm   (b) 249 nm   (c) 254 nm   (d) 259 nm

9. Why is the oxygen–hydrogen absorption of CH_3OH such a broad band in the infrared?

   (a) Rotational energy levels broaden the absorption.

   (b) Hyperconjugation resonance broadens the absorption.

   (c) Resonance broadens the absorption.

   (d) Hydrogen bonding broadens the absorption.

10. The CMR spectrum of an unknown compound shows 4 absorptions and the PMR spectrum shows 4 absorptions. Which of the following compounds is the unknown compound?

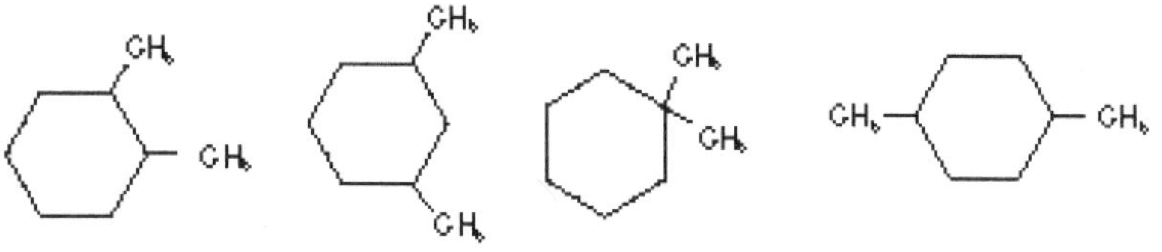

    (a)   (b)   (c)   (d)

11. The CMR spectrum of an unknown compound shows 6 absorptions and the PMR spectrum shows 5 absorptions. Which of the following compounds is the unknown compound?

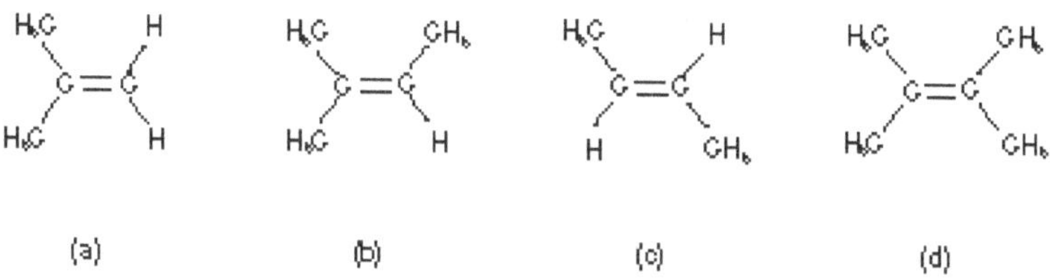

12. Which of the following alkenes would have the largest λ_{max}?

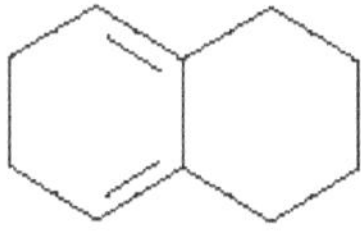

13. What is the λ_{max} for the following compound? Use the provided parameters for your calculation.

transoid base value = 214 nm
cisoid base value = 253 nm
alkyl groups on base = +5 nm
exocyclic C=C = +5 nm
extended conjugation = +30 nm

(a) 234 nm   (b) 244 nm   (c) 273 nm   (d) 283 nm

14. Which one of the following statements about ionization in a mass spectrometer is incorrect?

(a)   Gaseous atoms are ionized by bombarding them with high-energy electrons.

(b)   Atoms are ionized, so they can be accelerated.

(c)   Atoms are ionized, so they can be deflected.

(d)   It does not matter how much energy you use to ionize the atoms.

15. The path of ions after deflection depends on

    (a) only the mass of the ion

    (b) only the charge on the ion

    (c) both the charge and the mass of the ion

    (d) neither the charge nor the mass of the ion

16. Which of the following species will be deflected to the greatest extent?

    (a) $^{37}Na^+$ (b) $^{35}Na^+$ (c) ^{37}Na (d) $^{35}Na^{2+}$

17. Which of the following statements about tetramethylsilane is incorrect?

    (a) It produces a single peak at δ = 10.

    (b) It is inert.

    (c) It is volatile and can be easily distilled off and used again.

    (d) It is used to provide a reference against which other peaks are measured.

18. The proton NMR of 2-bromo-2-methylpropane will consist of

    (a) three quartets and a singlet   (b) two doublets and a singlet

    (c) two singlets   (d) one singlet

19. Which one of the following statements about protons on O–H groups is incorrect?

    (a) They always produce a doublet.

    (b) They always produce a singlet.

    (c) They are not taken into account when working out splitting patterns of adjacent protons.

    (d) Their signal can be removed from its normal location by shaking with deuterium oxide.

20. To work out the molecular mass of an organic molecule, one would look at its

    (a) infrared spectrum   (b) mass spectrum

    (c) proton NMR spectrum   (d) boiling point

21. The isomer of C_4H_8 which produces an NMR spectrum with four different signals is

    (a) $CH_2{=}CHCH_2CH_3$ (b) $CH_3CH{=}CHCH_3$

    (c) $(CH_3)_2C{=}CH_2$ (d) cyclobutane

22. Which one of the following hydrocarbons produces an NMR spectrum with more than one peak?

    (a) methane   (b) ethane

    (c) butane   (d) cyclobutane

23. Which compound has a molecular ion at $m/z = 58$, an infrared absorption at 1650 cm^{-1} and just one singlet in its NMR spectrum?

    (a) butane

    (b) CH_3COCH_3

    (c) CH_3CH_2CHO

    (d) 2-methylpropane

24. The splitting pattern for a signal is found by...

    (a) counting the number of chemically equivalent hydrogen atoms on adjacent atoms

    (b) counting the number of chemically different hydrogen atoms on adjacent atoms

    (c) counting the number of chemically different hydrogen atoms on adjacent atoms and adding 1

    (d) counting the number of chemically different hydrogen atoms on adjacent atoms and subtracting 1

25. The proton NMR of 2-bromopropane will consist of...

    (a) two doublets and a sextet

    (b) a doublet and a septet

    (c) a singlet, a doublet and a triplet

    (d) a singlet, a doublet and a triplet

26. Which of the following is not a use for mass spectrometry?

    (a) calculating the isotopic abundance in elements

    (b) investigating the elemental composition of planets

    (c) confirming the presence of O—H and C   O in organic compounds

    (d) calculating the molecular mass of organic compounds

27. Which one of the following statements about the mass spectrum of CH_3Br is correct?

    (a) The last two peaks are of equal size and occur at m/z values of 94 and 96.

    (b) The last two peaks have abundances in the ratio 3:1 and occur at m/z values of 94 and 96.

    (c) There is just one peak for the molecular ion with an m/z value of 95.

    (d) There is just one peak for the molecular ion with an m/z value of 44.

28. Which one of the following pieces of information cannot be obtained from an infrared spectrum?

    (a) the molecular mass

    (b) the presence of C   O bonds

    (c) the presence of O—H bonds

    (d) the identity of a compound through comparison with other spectra

29. To check that a secondary alcohol has been completely oxidized to a ketone you can

    (a) check that the IR spectrum has absorptions at 3500 cm^{-1} and 1650 cm^{-1}

    (b) check that the IR spectrum has no absorption around 3500 cm^{-1}.

    (c) check that the IR spectrum has no absorption around 1650 cm^{-1}

    (d) check that the IR spectrum has no absorptions at 3500 cm^{-1} and 1650 cm^{-1}

30. Signals in a proton NMR spectrum do not provide information about

    (a) the relative number of hydrogen atoms in a particular environment

    (b) the number of chemically different hydrogen atoms on adjacent atoms

    (c) the environment of different hydrogen atoms in a molecule

    (d) the molecular mass of an organic molecule

31. The proton NMR spectrum of propane will consist of...

    (a) a triplet and a singlet   (b) a triplet and a quartet

    (c) a doublet and a sextet   (d) a triplet and a septet

32. The proton NMR of 1-bromopropane will consist of...

    (a) two doublets and a sextet   (b) a doublet and a septet

    (c) a singlet, a doublet and a triplet   (d) two triplets and a sextet

33. Four major spectroscopic tools are listed below. Which makes use of the longest wavelength radiation?

    (a) infrared   (b) ultraviolet

    (c) visible   (d) proton NMR

34. You have three dyes. One is green, one is blue and one is yellow. Which absorbs the shortest wavelength of visible light, and which absorbs the longest?

    (a) longest = yellow; shortest = blue

    (b) longest = blue; shortest = green

    (c) longest = yellow; shortest = blue

    (d) longest = green; shortest = yellow

35. Which if any, of the following compounds will display spin–spin splitting in the ^{1}H NMR?

    (a) $(CH_3)_3COCH_3$ (b) $Br(CH_2)_3Br$

    (c) *para*-xylene, $CH_3C_6H_4CH_3$ (d) none of these

36. The ^{1}H NMR of 1,1-dibromoethane consists of two well-separated signals, one large and the other small. Which of the following descriptions is correct?

    (a) The large signal is a quartet and the small signal is a doublet.

    (b) The large signal is a triplet and the small signal is a singlet.

    (c) The large signal is a singlet and the small signal is a triplet.

    (d) The large signal is a doublet and the small signal is a quartet.

37. Which spectroscopic tool would be best for distinguishing a sample of 1,2,2-trichloropropane from 1,1,2-trichloropropane?

    (a) ^{1}H NMR   (b) infrared spectroscopy

    (c) ultraviolet–visible spectroscopy   (d) mass spectrometry

38. Which spectroscopic tool would be best for distinguishing a sample of chlorocyclopentane from bromocyclopentane?

    (a) ^{1}H NMR   (b) infrared spectroscopy

    (c) ultraviolet–visible spectroscopy   (d) mass spectrometry

39. Which spectroscopic tool would be best for distinguishing a sample of 1,3-cyclohexadiene from 1,4-cyclohexadiene?

    (a) ^{1}H NMR   (b) infrared spectroscopy

    (c) ultraviolet–visible spectroscopy   (d) mass spectrometry

40. Combustion analysis of an organic compound shows it to be 64.3% carbon. It displays a molecular ion at $m/z = 112$ amu in the mass spectrum. Which of the following ia a plausible molecular formula for this compound?

    (a) C_8H_{16} (b) $C_7H_{12}O$ (c) $C_6H_8O_2$ (d) $C_5H_4O_3$

41. Which type of C—H has the highest stretching frequency in the infrared spectrum?

    (a) $RCHO$ (b) RCH_3 (c) R_2C CH_2 (d) RC CH

42. Which C=O function has the lowest stretching frequency in the infrared spectrum?

    (a) acyl chloride   (b) aldehyde   (c) amide   (d) ester

43. Which hydrocarbon gives the lowest field ^{1}H NMR signal?

    (a) cyclohexane   (b) benzene

    (c) 1,4-cyclohexadiene   (d) 1-butyne

44. A C_2H_2BrCl compound gives a ^1H NMR spectrum consisting of two equal-sized doublets, J = 16 Hz. What is this compound?

    (a) (Z)-1-bromo-2-chloroethene   (b) (E)-1-bromo-2-chloroethene

    (c) 1-bromo-1-chloroethene   (d) none of the above

45. The ^1H NMR spectrum of diethyl ether shows

    (a) two peaks, one a triplet, the other a quartet

    (b) two peaks, one a triplet, the other a doublet

    (c) four peaks, all doublets

    (d) four peaks, all triplets

46. Compared with a conjugated diene, the UV–visible absorption spectrum of a conjugated triene will change in which way?

    (a) The λ_{max} will increase and the ε will decrease.

    (b) The λ_{max} will decrease and the ε will increase.

    (c) Both the λ_{max} and the ε will decrease.

    (d) Both the λ_{max} and the ε will increase.

47. Consider four $C_3H_5Cl_3$ isomers. Which has two ^1H NMR singlets and three ^{13}C NMR signals?

    (a) $CH_3CH_2CCl_3$ (b) $CH_2ClCHClCH_2Cl$

    (c) $CH_3CHClCHCl_2$ (d) $CH_3CCl_2CH_2Cl$

48. An unknown compound has a molecular ion at m/z = 79 amu in its mass spectrum. Analysis shows its composition to be 17.7% nitrogen. What is the molecular formula of this compound?

    (a) C_5H_5N (b) $C_4H_3N_2$ (c) C_3HN_3 (d) $C_4H_{17}N$

49. Neopentyl chloride, $(CH_3)_3CCH_2Cl$, reacts with the strong base sodium amide to form a new compound. This compound has a molecular ion at m/z = 70 amu, and displays two ^1H NMR singlets at δ 0.20 and 1.05 ppm (intensity ratio = 2:3) What is a plausible structure for this compound?

    (a) 2-methyl-2-butene   (b) 1,1-dimethylcyclopropane

    (c) methylcyclobutane   (d) cyclopentane

50. A compound has a molecular ion at m/z = 142 amu, and displays only one ^{1}H NMR signal. Which of the following satisfies these facts?

    (a)   methyl iodide

    (b)   1,1,2,2-tetrafluorocyclopentane

    (c)   *para*-disulphydrylbenzene [$C_6H_4(SH)_2$]

    (d)   2,4-hexadyne-1,6-diol

51. Which of the following statements is the best definition of the **base peak** in a mass spectrum?

    (a)   the molecular ion peak

    (b)   the lowest m/z peak

    (c)   the highest mass rearrangement ion

    (d)   the ion peak of greatest intensity

52. Four methyl compounds are listed below. Which has the lowest field methyl resonance in the ^{1}H NMR spectrum?

    (a) $(CH_3)_4Si$ (b) $(CH_3)_3N$ (c) $(CH_3)_2S$ (d) $(CH_3)_2O$

53. The following hydrocarbons all have ^{1}H NMR spectra consisting of a single sharp peak. Which exhibits the greatest shielding?

    (a) 2-butyne   (b) benzene

    (c) 1,2,3-butatriene   (d) 1,3-butadyne

55. A $C_5H_{12}O_2$ compound has strong infrared absorption at 3300 to 3400 cm^{-1}. The ^{1}H NMR spectrum has three singlets at δ 0.9 , δ 3.45 and δ 3.2 ppm; relative areas 3:2:1. Addition of D_2O to the sample eliminates the lower field signal. The ^{13}C NMR spectrum shows three signals all at higher field than δ 100 ppm. Which of the following compounds best fits this data?

    (a) 1,5-pentanediol   (b) 1,3-dimethoxypropane

    (c) 2,2-dimethyl-1,3-propanediol   (d) 2,4-pentanediol

56. Which of the following statements is not correct?

    (a)   Frequencies in cm^{-1} are much smaller numbers than frequencies in Hz.

    (b)   Wavelengths in μ are smaller numbers than wavelengths in Å.

    (c)   Frequency varies inversely with wavelength.

    (d)   Wavelengths given in nm are larger numbers than wavelengths in Å.

57. Which compound absorbs at the highest field in the ^{1}H NMR spectrum?

    (a) CH_3Cl (b) CH_2Cl_2 (c) CH_4 (d) $CHCl_3$

58. Which of these compounds absorbs at the highest field in the ^{1}H NMR spectrum?

    (a) CH_2Cl_2 (b) $CHCl_3$ (c) CH_3Cl (d) CH_3I

59. Which compound has the C-X vibration that occurs at the highest frequency in the IR spectrum?

    (a) CH_3CH_2Br (b) CH_3CH_2Cl (c) CH_3CH_2F (d) CH_3CH_2I

60. How many different types of carbon signals would be observed by ^{13}C NMR?

    (a) 1   (b) 2   (c) 3   (d) 4

Answers

1. (d)	2. (b)	3. (c)	4. (c)	5. (a)
6. (b)	7. (c)	8. (d)	9. (d)	10. (a)
11. (b)	12. (d)	13. (d)	14. (d)	15. (c)
16. (d)	17. (a)	18. (d)	19. (a)	20. (b)
21. (a)	22. (c)	23. (b)	24. (c)	25. (b)
26. (c)	27. (a)	28. (a)	29. (b)	30. (d)
31. (d)	32. (d)	33. (d)	34. (d)	35. (b)
36. (d)	37. (a)	38. (d)	39. (c)	40. (c)
41. (d)	42. (c)	43. (b)	44. (b)	45. (a)
46. (d)	47. (d)	48. (a)	49. (b)	50. (a)
51. (d)	52. (d)	53. (a)	54. (a)	55. (c)
56. (d)	57. (c)	58. (d)	59. (c)	60. (b)

GLOSSARY

Absorbance (A) It is defined as $A = \log(I_o/I)$, where I_o is the radiant power of light striking a sample on one side and I is the radiant power emerging from the other side.

Absorption filter A coloured medium that transmits a relatively narrow portion of the electromagnetic radiation.

Absorption spectrum A graph of absorbance or transmittance of light as a function of wavelength, frequency or wavenumber.

Aspiration The process by which a sample solution is drawn by suction into the fuel/oxidant mixture.

Asymmetric stretching In these vibrations, one atom approaches the central atom while the other departs from it.

Atomic absorption spectroscopy A technique in which the absorption of light by gaseous atoms in a flame or furnace is used to measure the concentration of atoms.

Atomic emission spectroscopy A technique in which the emission of light by thermally excited atoms in a flame or furnace is used to measure the concentration of atoms.

Atomic fluorescence Radiant emission from atoms that have been excited by electromagnetic radiation, the energy of which exactly matches differences in levels of the atomic energy states.

Atomization The process of generating an atomic gas by applying heat to the sample.

Attenuator A device for diminishing the radiant power in the beam of an optical instrument.

Auger electron spectroscopy A surface analysis technique based on the energy analysis of Auger electrons typically excited by a focused electron beam.

Band spectra Regions in which spectral lines are so numerous and close together that they cannot be resolved.

Base peak A peak corresponding to the ion of maximum abundance. It is the most intense peak.

Bathochromic shift A shift in the wavelength to longer wavelengths (red shift).

Beamsplitter A device for dividing radiation from a monochromator such that one portion passes through the sample while the other passes through the blank.

Chemical interference In general, a substance that prevents accurate determination of the analyte. In atomic absorption, emission, or fluorescence spectrometry, a sample constituent that acts as a multiplicative interference by decreasing the efficiency with which the analyte element is converted to gas-phase atoms.

Chemical isomer shift (in Mössbauer spectroscopy) The shift that arises due to the non-zero volume of the nucleus and the electron charge density due to s-electrons within it.

Chemical shift A variation in the resonance frequency of a nuclear spin due to the chemical environment around the nucleus. Chemical shift is reported in ppm.

Chemical shift anisotropy In nuclear magnetic resonance, the directional dependence of electronic shielding in a molecule.

Chopper A mechanical device that alternately transmits and blocks radiation from a source.

Continuous wave (CW) A form of spectroscopy in which a constant-amplitude electromagnetic wave is applied.

Correlation spectroscopy (COSY) A term for any two-dimensional NMR experiment that produces signals that correlate two nuclei by their chemical shifts. It has also come to mean the proton–proton chemical shift correlation experiment.

Cuvette The container that holds the analyte in the light path (absorption spectroscopy).

Deshielding of nucleus When the effective magnetic field experienced by the nucleus is more than the applied field, the nucleus is said to be deshielded.

Deuterium lamp A source that provides continuous radiation in the ultraviolet region of the spectrum.

Diffraction grating A common dispersive element for the monochromator. It consists of equally spaced grooves positioned on a metal surface and can disperse the radiation into component energy bands of different wavelengths by rotating the grating relative to the radiation source.

Diode array detector A silicon chip that accommodates numerous photodiodes, and provides the capability to scan entire spectral regions simultaneously.

Double-beam spectrophotometer An instrument in which the beam is split to permit comparison of sample and solvent at the same time. Operation is usually automated.

Dynode An intermediate electrode in a photomultiplier tube.

Electromagnetic radiation A form of energy with properties that can be described in terms of waves or alternatively as particulate photons, depending on the method of observation.

Electron paramagnetic spectroscopy (EPR) Another name for electron spin resonance spectroscopy.

Electronic spectroscopy Absorption or emission of radiation, usually in the visible and UV region, accompanying changes in electronic state.

Electron spectroscopy for chemical analysis (ESCA) Another name for X-ray photoelectron spectroscopy.

Electron spin resonance spectroscopy (ESR) Observation of transitions between energy levels associated with unpaired electrons in magnetic field.

Electronic transition The promotion of an electron from an electronic state to an excited electronic state and conversely.

Fluorescence The process in which a molecule emits a photon 10^{-8} to 10^{-4} s after absorbing photon.

Fourier transform (FT) A mathematical technique capable of converting a time domain signal to a frequency domain signal and vice versa.

Franck Condon principle An electronic transition takes place so rapidly that a vibrating molecule does not change its internuclear distance appreciably during the transition.

Fundamental vibrations It corresponds to the first vibrational transition from zeroth vibrational level to the first vibrational level.

Group frequencies Frequencies in absorption spectra, particularly in the infrared region, associated with specific functional groups.

Gyromagnetic ratio The ratio of the resonance frequency to the magnetic field strength for a given nucleus.

Hollow cathode lamp One that emits sharp atomic lines characteristic of the element from which the cathode is made.

Hyperchromic shift An effect due to which the intensity of absorption maximum increases.

Hypochromic shift An effect due to which the intensity of absorption maximum decreases.

Hypsochromic shift An effect by virtue of which the absorption maximum is shifted towards shorter wavelength. Also known as blue shift.

Infrared region A portion of the electromagnetic spectrum from 1 μm to 1 mm in wavelength. It is from 4000 to 667 cm^{-1} for structural elucidation purposes. The high-frequency region 4000–1300 cm^{-1} is called functional group region and 1300–667 cm^{-1} region is called the fingerprint region.

Interference filter An optical filter that provides narrow bandwidths owing to constructive interference.

Larmor frequency The resonance frequency of a spin in a magnetic field; the rate of precession of a spin packet in a magnetic field; the frequency which will cause a transition between the two spin energy levels of a nucleus.

Magic angle spinning Technique for narrowing NMR lines due to solids.

Mass spectrometer A device that ionizes and often fragments gaseous molecules and then separates them by mass and measures the quantity of each ion.

Mass spectrometry Analysis by mass and charge of molecular ions and their breakdown products. Often prefixed by a term describing means of obtaining ions.

Matrix The medium that contains the analyte.

McLafferty rearrangement In mass spectrometry, it refers to beta cleavage rearrangement of organic ions.

Monochromatic radiation Ideally, electromagnetic radiation that consists of a single wavelength.

Mössbauer spectroscopy Absorption of nuclear gamma rays by nuclei. Gives information about chemical environment of and magnetic interactions and electric field gradient at the nuclei.

Nebulization The process of breaking a liquid into a mist of fine droplets.

Nernst glower A source of infrared radiation.

Nichrome A nickel/chromium alloy; when incandescent, a source of infrared radiation.

Nitrogen rule A molecule of even-numbered molecular mass must contain no nitrogen atom or an even number of nitrogen atoms. An odd-numbered molecular mass requires an odd number of nitrogen atoms.

Nuclear Overhauser effect (NOE) An effect that occurs between nuclei that are very close in space to one another. The NOE is used for determining internuclear distances and for the enhancement of signals from insensitive nuclei.

Overtones The term overtone in IR spectroscopy is used to indicate a multiple of fundamental frequency.

Phosphorescence Emission process in which the upper and lower states have different spin multiplicities. In organic molecules, the upper state is usually the lowest triplet state and the lower state is the ground state.

Photomultiplier tube A sensitive detector for electromagnetic radiation; amplification is accomplished by a series of dynodes.

Phototube A common photoelectric detector in the UV-visible and near IR regions, particularly in less expensive instruments.

Precessional frequency It is equal to the number of revolutions per second made by the magnetic moment vector of the nucleus around the external field.

Prism An optical body that disperses polychromatic radiation into its component wavelengths.

Raman spectroscopy It involves the study of vibrational–rotational energy changes in molecules by means of scattering of light.

Rayleigh scattering It is defined as the scattering of radiation without any change in frequency.

Resonance Raman spectroscopy Raman spectroscopy performed with an irradiation frequency that is absorbed by sample.

Rocking Type of bending vibration in which the movement of atoms takes place in the same direction.

Rule of mutual exclusion principle In a molecule, which has a centre of symmetry, the Raman active vibrations are infrared-inactive and vice versa.

Scissoring In this type of bending, two atoms approach each other.

Shielding of nucleus When the effective magnetic field experienced by the nucleus is less than that of the applied field, the nucleus is said to be shielded.

Single beam spectrophotometer An instrument with one optical path in which the sample and solvent are examined separately, usually operated manually.

Spectrophotometer An instrument designed to measure transmittance or absorbance of samples and a prism or grating monochromator is used as the wavelength isolation device.

Sputtering The process whereby an atomic cloud is produced by collisions between excited argon ions and the element on the surface of the cathode of a hollow cathode lamp.

Stokes lines When the scattering of radiation produces lines with lower frequency compared to that of the incident beam.

Symmetric stretching In this type of stretching vibration, the movement of the atoms with respect to a particular atom in a molecule is in the same direction.

Tungsten filament lamp A convenient source of visible and near-infrared radiation.

Twisting In this bending vibration, one of the atoms moves up the plane while the other moves down the plane with respect to the central atoms.

Vibrational transition Transitions between vibrational states of the ground electronic state that are responsible for infrared absorption.

Visible radiation That portion of the electromagnetic radiation (340–780 nm) to which the human eye is perceptive.

Wagging Bending vibration, in which two atoms move up and down the plane with respect to the central atom.

Wavelength of electromagnetic radiation The distance between successive maxima or minima of a wave.

X-ray photoelectron spectroscopy Velocity analysis of photoelectrons emitted from sample irradiated by X-ray source.

Zeeman splitting Dipolar interaction with the magnetic field experienced by the nuclear spin moment in the presence of the magnetic field.

REFERENCES

Christian, G.D. and O'Reily, J.E. *Instrumental Analysis*, 2nd (edn.). McGraw-Hill, New York. 1987.

Daniel, C.Harris. *Exploring Chemical Analysis*.W.H.Freeman & Company, New York 1997.

Douglas, A. Skoog, Donald, M.West and James Holler,F. *Fundamental of Analytical Chemistry*, 7th edn. Saunders College Publishing, New York.1996.

Ebdon, L., Evans, E.H., Fisher, A. and Hill, S.J. *Analytical Atomic Spectrometry*, 2nd edn. VCH, New York. 1998.

Ebsworth, E.A.V., David,W.H., Rankin and Stephen Cradock. *Structural Methods in Inorganic Chemistry*, Blackwell Scientific Publications, Oxford ELBS 1988.

Frank, A.Settle.(ed.). *Handbook of Instrumental Techniques for Analytical Chemistry*, Prentice Hall, Upper Saddle River, NJ.1997.

Gardiner, D.J. and Graves, P.R. (eds.). *Practical Raman Spectroscopy*. Springer-Verlag, Heidelberg.1989.

Hanswell, S.J. (ed.). *Atomic Absorption Spectrometry*. Elsevier, Amsterdam. 1992.

Jenjins Ron. *An Introduction to X-Ray Spectrometry*. Heyden & Sons, London. 1974.

Johnstone, R.A.W.and Rose, M.E. *Mass Spectrometry for Chemists and Biochemists*. Cambridge University Press, Cambridge.1996.

Metcalfe, E.*Atomic Absorption and Emission Spectroscopy*.John Wiley & Sons,Chichester.1987.

Olsen, E.D. *Modern Optical Methods of Analysis*. McGraw-Hill, New York. 1975.

Rubinson, K.A. and Rubinson, J.F. *Contemporary Instrumental Analysis*. Prentice Hall, Englewood Cliffs, NJ. 2000.

Sanders, J. and Hunter, B. *Modern NMR Spectroscopy: A Guide for Chemists*, 2nd edn. Oxford University Press, Oxford.1993.

Silverstein, R.M., Bassler,G.C. and Morril, T.C. *Spectrometric Identification of Organic Compound*, 5th edn. John Wiley & Sons, New York. 1991.

Skoog, D.A. and Leary, J.J. *Principles of Instrumental Analysis*, 4th edn. Saunders College Publishing, Philadelphia. 1992.

Strobel, H. and Heineman, W.R. *Chemical Instrumentation: A Systematic Approach*, 3rd edn. Addison-Wesley, Boston. 1989.

Thomas, M.J.K. and Ando, D.J. *Ultraviolet and Visible Spectroscopy*. John Wiley & Sons, Chichester. 1996.

Wertz, J.E. and Bolton, J.R. *Electron Spin Resonance, Elementary Theory and Practical Applications*. McGraw-Hill, New York.1972.

Willard, H.H. and *et al. Instrumental Methods of Analysis*,7th edn.Wadsworth, Belmont,CA.1987.

INDEX

www.ingramcontent.com/pod-product-compliance
Lightning Source LLC
LaVergne TN
LVHW051255200726
843510LV00010B/1125